Sustainable Agricultural Value Chain

Edited by Habtamu Alem and Pradyot Ranjan Jena

Published in London, United Kingdom

IntechOpen

Supporting open minds since 2005

Sustainable Agricultural Value Chain
http://dx.doi.org/10.5772/intechopen.94780
Edited by Habtamu Alem and Pradyot Ranjan Jena

Contributors
Azwihangwisi E. Nesamvuni, James Bokosi, Khathutshelo A. Tshikolomo, Ndivhudzannyi S. Mpandeli,
Cebisa Noxolo Nesamvuni, Michael Paul Kramer, Jon H. Hanf, Linda Bitsch, Marco Nuti, Maria Cristina
Echeverria, Sania Ortega-Andrade, Sebastián Obando, Eduardo Eugênio Spers, Luciana Florêncio de
Almeida, Laleska Rossi Moda, Sandra Mara de Alencar Schiavi

Notice
Statements and opinions expressed in the chapters are these of the individual contributors and not
necessarily those of the editors or publisher. No responsibility is accepted for the accuracy of
information contained in the published chapters. The publisher assumes no responsibility for any
damage or injury to persons or property arising out of the use of any materials, instructions, methods
or ideas contained in the book.

First published in London, United Kingdom, 2022 by IntechOpen
IntechOpen is the global imprint of INTECHOPEN LIMITED, registered in England and Wales,
registration number: 11086078, 5 Princes Gate Court, London, SW7 2QJ, United Kingdom
Printed in Croatia

British Library Cataloguing-in-Publication Data
A catalogue record for this book is available from the British Library

Additional hard and PDF copies can be obtained from orders@intechopen.com

Sustainable Agricultural Value Chain
Edited by Habtamu Alem and Pradyot Ranjan Jena
p. cm.
Print ISBN 978-1-83969-755-5
Online ISBN 978-1-83969-756-2
eBook (PDF) ISBN 978-1-83969-757-9

We are IntechOpen,
the world's leading publisher of
Open Access books
Built by scientists, for scientists

6,100+
Open access books available

149,000+
International authors and editors

185M+
Downloads

Our authors are among the

156
Countries delivered to

Top 1%
most cited scientists

12.2%
Contributors from top 500 universities

CLARIVATE ANALYTICS

BOOK
CITATION
INDEX

INDEXED

WEB OF SCIENCE™

Selection of our books indexed in the Book Citation Index (BKCI)
in Web of Science Core Collection™

Interested in publishing with us?
Contact book.department@intechopen.com

Numbers displayed above are based on latest data collected.
For more information visit www.intechopen.com

Meet the editors

Dr. Habtamu Alem is a research scientist at the Department of Economics and Society at the Norwegian Institute of Bioeconomy Research (NIBIO). He has published papers in the *International Journal of Production Economics*, *International Journal of Educational Development*, *Journal of Sustainability*, *Journal of Applied Economics*, *Journal of Economies*, *International Journal of Productivity and Performance Management*, and *Journal of Research in Economics*, among others. He regularly reviews academic publications. He is currently editing a book titled *Public Economics: New Perspectives and Uncertainty, which will be published in 2023 by IntechOpen*

Dr. Pradyot Ranjan Jena is an associate professor at the School of Management, National Institute of Technology Karnataka (NITK), Surathkal in India. Dr. Jena's core research areas are impact evaluation development programs, food security, climate change adaptations in agriculture, and energy economics. He has led ten national and international research projects funded by IFAD, CGIAR, the Government of Australia, the Indian Council of Social Science Research, Leibniz University of Hannover, Germany, and the Poverty, Equity, and Growth Network (PEGNet). He has published over 50 research papers in Web of Science- and SCOPUS-indexed peer-reviewed journals, international conference proceedings, and book chapters. He has published two books. He serves on the editorial board of several reputed international journals including *Frontiers in Sustainable Food Systems and Frontiers in Climate Change*.

Contents

Preface

Our world relies heavily on value chains, but we are still learning how to make them sustainable. Increasingly competitive markets require farms to differentiate their products while strengthening their brand reputation. Sustainability standards have the potential to appeal to a growing number of customers who are concerned about the environment as well as social and ethical issues. This book advocates a value-chain approach to improving certification governance and standards. Monitoring the impact of climate change and mitigating it by providing irrigation to increase the quantity of value-added products are both critical. Meanwhile, blockchain technology has the potential to transform business models and supply chain networks in the agri-food industry. This book discusses scientific, technical, and societal challenges focused on the coffee value chain. Changes in public policies are required to conform with territorial development programs based on sustainable development objectives.

Habtamu Alem
Research Scientist in the Department of Economics and Society,
Norwegian Institute of Bioeconomy Research (NIBIO),
Ås, Norway

Pradyot Ranjan Jena
National Institute of Technology Karnataka (NITK),
Surathkal, India

Value Chains Constraints, Technology and Adoption

Chapter 1

Scientific, Technical, and Social Challenges of Coffee Rural Production in Ecuador

Echeverría María Cristina, Ortega-Andrade Sania, Obando Sebastián and Marco Nuti

Abstract

The production of coffee in Ecuador a family activity carried out in rural areas. Due to the economic importance of this crop and its ability to adapt to different ecosystems, it has been widely introduced in government conservation and economic reactivation programs. At the present, it is cultivated in the four Ecuadorian natural regions that comprise the Amazon rainforest, the Andean mountains, the Pacific coast, and the Galapagos Islands. The different climate and altitude characteristics of these regions allow Ecuador to grow all commercial varieties of coffee. The variety planted, the region of origin, and the type of post-harvest processing gives each cup of coffee a unique flavor and aroma. To recovery the knowledge behind each production process, a complete review of the whole coffee productive chain was made. The information reviewed was compared with the available information of other neighboring countries and complemented with experiences described by small farmers. The analysis confirms that Ecuador has a competitive advantage due to its ecosystem diversity. However, the development of this industry depends on the correct implementation of policies that cover three main aspects: (1) farmers' quality of life, (2) training and research programs, and (3) fair trade for small producers.

Keywords: coffee agroecosystems, coffee processing, coffee by-products, rural coffee production, organic coffee

1. Introduction

Coffee is one of the most popular and consumed beverages in the world. High coffee consumption can have a substantial effect on health [1]. It is among the most traded agricultural commodities. In 2020, it is estimated that 10,520,820 tons of coffee were produced [2] and almost the same amount was consumed [3]. In Latin America, its production is an integral component of the livelihoods of millions coffee farmers, associates, and workers including their families [4].

Coffee has been cultivated in Ecuador since the eighteenth century. It is one of the ten most important crops, being grown entirely in rural areas. In total, the coffee area exceeds 30 thousand hectares planted [5]. Due to the geographical characteristics of Ecuador, it is one of the few countries in the world that cultivate the two commercial varieties: *Coffea arabica* or arabica coffee and *Coffea canephora* or robusta coffee. In 2020, Ecuador exported 14,828.15 bags of 60 kilos [6].

IntechOpen

One of the advantages of coffee cultivation is its adaptability to different ecosystems, in which it produces important environmental benefits. In Ecuador, coffee trees are managed as agroforestry systems. According to each region and its climatic conditions, coffee is grown together with forest species, mainly fruit trees, which provide temporary shade to the crop, and timber species that provide permanent shade [7]. This landscape arrangement contributes to the maintenance of appropriate habitat for various species of flora and fauna, the capture of carbon in the soil, and the water balance of ecosystems [8].

Despite the environmental advantages offered by this crop, the productive sector is affected not only by the consequences of the pandemic and the deterioration of the global economy but also by the change in climatic conditions that promote the migration and spread of pathogenic organisms such as the coffee leaf rust [9] and the coffee cherry borer which is very difficult to eradicate [10]. They are not the only plagues that affect coffee cultivation, but they are the ones that have caused the greatest economic losses in coffee production around the world [11].

The combination of strategies such as the use of chemical fungicides, quarantines, cultural practices, biocontrol agents, and the selection of resistant varieties have helped to reduce pests and diseases. However, climate change is threatening the survival of *Coffea arabica* cultivation worldwide [12]. This pushed the producers to review the agronomic practices and search for new strategies. Among the different proposed options, forest conservation seems to be the most promising to guarantee coffee production in the future. The forests are a source of water, help to mitigate the global temperature rise, and are a source of microorganisms that could be used to regenerate eroded soils and counteract pathogenic organisms [13, 14].

Within this context, it is evident that Ecuador has optimal geographic and environmental conditions to produce quality coffee and overcome the new challenges of climate change. However, its production is lower compared with neighboring productive countries like Peru, Colombia, and Brazil. This fact is associated with other problems that are limiting the development of this productive sector. In this review, social, technical, and scientific aspects are analyzed in the whole production chain to understand the opportunities, needs of coffee growers, and the limitations in the production process. The importance of this research lies in unifying the information of the different regions and coffee productive associations that are scattered throughout the national territory and rescue the needs of small farmers often not considered.

The next sections will describe the production of coffee in rural areas, its challenges and opportunities, the development policies, the social and economic importance, and a description of the production process from planting to waste management. The tables and graphs collect important information about coffee growers-associations, crops distribution, cultivated varieties, and their characteristics. The photographs included in this chapter show images of a typical coffee farm in the Ecuadorian Andes.

2. Rural production of coffee in Ecuador: challenges and opportunities

In Ecuador, coffee farming is an activity that has been passed down from generation to generation. It is carried out entirely in rural areas, which are characterized by having very productive soils but a high rate of poverty, low percentage of basic education, absence of basic services, and bad connecting roads.

The first records of coffee export in Ecuador date back to 1980 in the province of Portoviejo Manabí [15]. Subsequently, thanks to the opening of world trade and the adaptability of this plant in the different Ecuadorian ecosystems, coffee is currently grown in 23 of the 24 provinces, becoming one of the 10 most important crops from an economic point of view [16]. Despite its importance, coffee production has been marked by ups and downs [17] and has substantially failed in improving the living conditions of rural farmers [18].

The crisis in the coffee sector covers many social and technological aspects and shows the little success of development policies. For example, the current Ecuadorian policy, in its 2015–2025 proposal, promote Sustainable Development Goals (SDGs). Literally, it suggests sustainable rural territorial development through the empowerment of the seven territorial planning areas that cover the entire country [19]. Even though, this policy has not been able to stop internal migration. According to projections by the National Institute of Statistics and Censuses (INEC), until 2010, only 35% of the population lived in rural areas with an annual decrease of 1.3% [20]. These data explain the aging of farmers because of young migration in search of better opportunities.

In the same way, the legal framework and the Constitution of Ecuador (2008) consider small farmers as priority groups for development. According to data from the FAO, more than 64% of the Ecuadorian agricultural production is in the hands of small producers categorized as Rural Farming Families (RFFs), on whom internal consumption depends. The RFFs represents 84.5% of the Agricultural Production Units (APU) [20]. These data highlight the importance of the rural sector in food production and the urgent need to change strategies to boost sustainable agriculture development. Unfortunately, not all NGOs development programs meet the needs of the population because they often replicate models used elsewhere resulting in a lack of cooperation between farmers.

In another context, Ecuador is considered a megadiverse country [21] and must find a balance between development and biodiversity conservation. With this goal, the Ministry of the Environment in 2011 proposed to increase protected areas by reducing the rate of deforestation, remedying environmental liabilities, reducing the use of pesticides, and addressing climate change through sustainable policies. To achieve part of this objective, the "Socio-Forest Program" and the "National Forestation and Reforestation Program" were created [22, 23]. However, there were many contradictions in the application of these measures. As an example, in the Amazon province of Orellana, palm crops (*Elaeis guineensis*) are still in constant and extensive growth, because of government subsidies. Currently, there are large private monocultures that displace other species, favoring the frontier expansion of agriculture and the loss of biodiversity [24, 25]. In the agricultural area, the government subsidizes and encourages the acquisition of seeds and the use of pesticides without appropriate control and monitoring.

For cocoa (*Theobroma cacao*) and coffee, the situation was more favorable, although some of the inconsistencies remain. These two crops form agroecosystems that, due to their structure and function, help to maintain a diversity of birds, bats, non-flying mammals, and invertebrates which in turn help to contain pathogens spread [26, 27]. In addition, these crops were part of the government's economic reactivation program from 2012- to 2021 and were consolidated through the formation of the Coffee and Cocoa Coordination Unit to promote productivity and forest conservation [28]. The coffee-cocoa program achieved the goal of promoting its cultivation, increasing its production in rural areas. However, much remains to be done in terms of quality. Rural farmers put all their resources into this crop motivated by government incentives, nonetheless, many of them express a feeling of abandonment and expect greater support that allows them to recover the investment and get out of poverty.

In conclusion, the policies of economic reactivation and conservation of rural areas must be reconsidered and oriented to grant a good quality of life for farmers. It is the only way to ensure sustainable development.

3. Social and economic importance of coffee cultivation

Coffee cultivation covers about 14% of the agricultural area of the country [29]. It is known as the unit crop due to its extension throughout the territory and the inclusion of all indigenous communities such as Quichua in the Andean region, Tsáchila in the Coastal region, and Shuar in the Amazon region.

Coffee production has a growing world demand and generates rural and urban employment because field activities include those necessary for the commercialization, transport, and industrialization processes. Also, it generates foreign income due to exportation. Export earnings are estimated to be between 60 and 80 billion dollars per year [30]. This income indicates the economic importance of the coffee sector in Ecuador. However, farmers are not the main beneficiaries of the coffee industry. Most of the income remains with the intermediaries who sell coffee on the international market [30]. An indicator of this reality is the high poverty rates in rural areas, which exceed 42% [20].

To survive, small farmers had to diversify their cropland. A small part is used to grow short-cycle food for sale and self-consumption. They also raise animals to obtain more economic income [29]. The difficult economic situation explains the migration of the younger population to the cities in search of better economic opportunities. This migration has caused the aging of farmers. It is estimated that the average age of coffee farmers is 50 years [20, 29, 31]. This means that over time their work capacity will decrease and there will be no new generations that continue with the activity.

To rescue coffee cultivation, small producers created the "benefits" that are legal associations that ensure fair trade. Unfortunately, 95% of coffee producers do not belong to any association [29]. One explanation for this fact is that most of the farms are in places of difficult access and do not have good communication channels. Phone signals do not work, and they live in isolation. Improving access to roads and basic services should be the government's priority to improve productivity. **Table 1** shows the main coffee growers' associations in Ecuador.

Coffee growers Associations	Province of action
Asociación Nacional Ecuatoriana de café (ANECAFÉ)	All provinces
Federación Regional de Asociaciones de Pequeños Cafetaleros Ecológicos del Sur (FAPECAFE)	Loja, El Oro y Samora Chinchipe
Asociación Agroartesanal de Caficultores "Río Íntag" (AACRI)	Imbabura y Pichincha
Asociacion De Productores Y Comercializadores De Cafe Organico Bosque Nublado Golondrinas.	Carchi
Empresa de Comercialización Asociativa de Manabí (COREMANABA) Corporación Ecuatoriana de Cafetaleros (CORECAF) Federación de Asociaciones Artesanales de Producción Cafetalera Ecológica de Manabí (FECAFEM).	Portoviejo, Guayas
Asociación Aroma amazónico	Sucumbios y Orellana
Asociación Aroma amazónico	Sucumbios y Orellana

Table 1.
Ecuadorian coffee growers association by Provinces.

Another important issue is the fact that farmers produce coffee, but they don't have the culture of coffee, understanding that they are not coffee drinkers. This is a fundamental difference from other neighboring countries such as Colombia where a true culture of coffee has been achieved and promoted through tourist activities that generates additional income for coffee growers. This data was observed after visiting several rural farms in the Andean region.

Creating a culture of coffee around this drink could influence the quality of the final product. The National Ecuadorian Coffee Association (ANECAFE) organizes different tasting events for this purpose.

4. Coffee production under agroforestry systems: farming practices, pest, and disease management

The quality of coffee depends on its organoleptic characteristics, which in turn depend on many other factors, including the genetics of the plant, the environmental conditions, the agricultural practices, the degree of cherry ripeness, and the post-harvest processing and the storage and transport conditions. Each step in coffee production is then of fundamental importance for obtaining the golden bean [32].

Due to the different ecosystem characteristics, Ecuador is one of the few countries in the world that can cultivate the two commercial species of coffee, *Coffea arabica* (70%) and *Coffea canephora* (30%). **Figure 1** shows the distribution of coffee crops by region and the cultivated areas in hectares. As can be seen, the cultivation of coffee is spread practically throughout the territory. On the other hand, **Table 2** shows the environmental characteristics of each region that directly influences the type of variety cultivated and the end quality of the final product [33, 34].

Of the two species, *C. arabica* is more appreciated for its organoleptic properties, and therefore it generates more revenue. The main characteristic of this variety is that it contains less caffeine compared with *C. canephora*. Caffeine is responsible for the bitter and strong taste [35].

Cultivated Coffee species by region in Ecuador

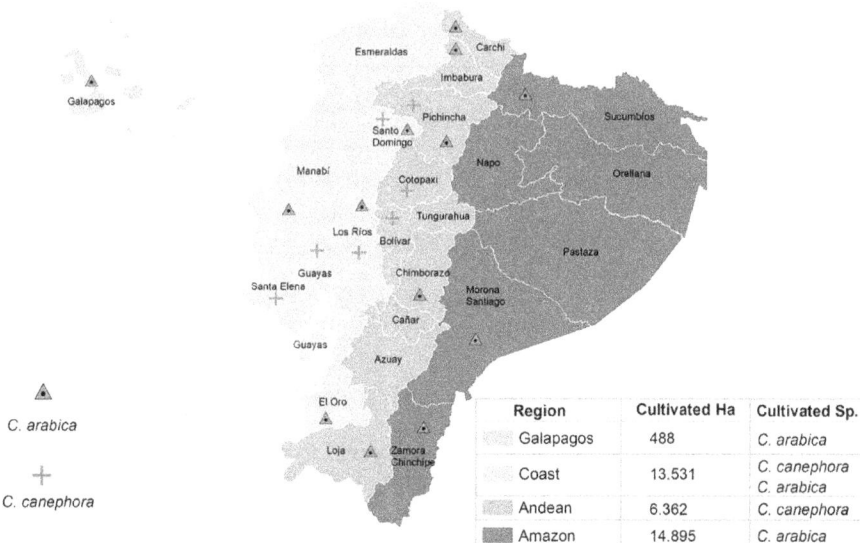

Region	Cultivated Ha	Cultivated Sp.
Galapagos	488	C. arabica
Coast	13.531	C. canephora / C. arabica
Andean	6.362	C. canephora
Amazon	14.895	C. arabica

Figure 1.
Distribution of Coffea arabica *and* Coffea canephora *in Ecuador.*

Region	Environmental conditions for coffee cultivation
Andean	18–23°C 1500–2900 masl
Coast	25–30°C 40–600 masl
Amazon	23° <500 masl
Galapagos	25°C 180–450 masl

Table 2.
Environmental conditions of coffee plantations by region.

    C. arabica usually grows in cooler climates common in high altitudes (18–21°C). For this reason, it is known as mountain coffee. Despite its adaptability to colder climates, Ecuador benefits from the Humboldt Current that provides fresh cold air during coffee flowering. It allows the cultivation of arabica varieties also in lower and warmer regions of the country. For example, Manabí on the coast grows arabica coffee at 600 meters above sea level (masl) [36], and in Galapagos, Santa Cruz, and San Cristobal Islands, a premium arabica coffee is obtained at altitudes between 180 and 450 masl [33]. The cold temperatures let a slow ripening of the cherry which allows a greater accumulation of sugars and metabolites that contribute to improving the organoleptic properties [37]. Global warming is therefore an enemy that puts arabica coffee production at risk [38]. One effect of this warming is related to the resentful presence of the coffee berry borer (*Hyphotenemus Hampei*) in higher and colder areas [10].

    The most cultivated varieties of *C. arabica* in Ecuador are Bourbon, Typica, Caturra, Geisha, Catucaí, Timor, Castillo Sachimor, and Sidra [39]. Choosing the right variety is the first important decision that agriculture must take. As reported in **Table 3**, some varieties seem to be more productive, resistant to pests, and more adaptable to high temperatures [40]. On the contrary, other varieties are less productive or resistant but have superior organoleptic characteristics. The genetics of the plant mostly determines the biochemical composition of the fruit (caffeine, sugars, lipids, and chlorogenic acids) and therefore the organoleptic properties [41]. For this reason, the decision of which variety to plant depends not only on the place of cultivation and climatic condition but also on the quality or volume of coffee that is desired to produce.

    Most coffee growers in Ecuador prefer the quality of the final product due to the higher economic income obtained with an excellent coffee rating. The National Ecuadorian Coffee Association (ANECAFE), for example, awards the best producers with the "golden cup". The golden cup is a contest in which international experts evaluated different coffee quality parameters like aroma, sweetness, body, color, and others. In 2021, the Sidra and Geisha varieties, grown in the Andean provinces of Pichincha and Loja, were the most appreciated by the expert's panel, achieving quality scores higher than 90 points [42]. This nomination allows farmers to sell coffee at prices 10–100 times higher.

    On the other hand, robusta coffee grows in warmer places on the coast and in the Amazon region. It is more productive, resistant to high temperatures and pests, and contributes mainly to the local market. It is appreciated in special mixes and soluble coffee production [43]. Maincrop differences between *C. arabica* and *C. Canephora* are summarized in **Table 4** [40].

Variety	Characteristics
Bourbon	It is originated from the Typica variety. It is known for its excellent cup quality. It has a 30% higher productivity than the typical variety. It reaches heights of 3 m, being susceptible to winds. Its maturation is early and there is a risk of fruit falling due to rain
Typica	Originally from Ethiopia, it began the history of coffee cultivation in America. It is characterized by being high (4 m). It has low productivity and is susceptible to rust. However, its cup is highly valued. It grows between 1300 and 1800 masl
Caturra	Arose from a mutation of bourbon. It is a low height (1.8 m). Its fruits can be red and yellow. They are characterized by early maturation. They tolerate drought, wind, and sun exposure better
Geisha	Originally from Ethiopia. The most important characteristic is its excellent cup and for that reason, it still occupies a place in production, however, it has low productivity and resistance to rust
Catucaí	It comes from an artificial cross between the Mundo Novo and Caturra varieties carried out in Brazil. They reach heights of 2.25 m and have high productivity (7.9 tons/hectare). The maturation of its fruit is late, being beneficial for areas where maturation coincides with the rainy season
Timor	The Timor Hybrid is originated from a spontaneous cross between the Typica variety of *C. arabica* and Robusta of *C. canephora*, identified around 1917 on the island of East Timor (Indian Ocean). Highly resistant to rust pathogen. The trait of genetic resistance comes from the species *C. canephora*
Castillo	It is originated from Colombia. It was developed by the National Coffee Research Center (Cenicafe). It is a pest-resistant plant characterized by being precocious and highly productive
Sidra	Discovered in Ecuador as a result of a cross between the Typica and Bourbon varieties. It is known for its aroma of flowers and fruits
Sarchimor	The Sarchimores are plants of low size, green or bronze bud, vigor, and high production, well adapted to low and medium altitude areas, and good cup quality

Table 3.
Most common C. arabica *Varieties cultivated in South and Central America.*

	Arabica	Robusta
Time from flower to ripe cherry	9 months	10–11 months
Yield (kg beans/ha)	1500–2000	2300–4000
Optimum temperature (yearly average)	15–24°C	24–30°C
Optimal rainfall	1500–2000 mm	2000–3000 mm
Optimal altitude	1000–2000 masl	0–700 masl
Caffeine content of beans	0.8–1.4%	1.7–4%

Table 4.
Maincrop differences between C. arabica *and* C. canephora.

In terms of genetics, there is significant variability of bean chemical composition and organoleptic characteristics between arabica and robusta and within variety levels [41]. Therefore, genetic gains for quality can be achieved by hybridization strategies and the use of new genomic tools that offers the opportunity to accurately decipher the genomic control of quality components.

Coffee is grown under agroforestry polyculture systems that, on the one hand, allow the conservation of biodiversity and, on the other hand, provide multiple advantages to coffee plantations. The trees provide shade, helping to maintain a suitable temperature. They also form a barrier that prevents damage from strong winds and rain and counteracts the spread of pathogens. In addition, they prevent soil erosion, forming ecosystem corridors that allow maintaining a considerable biodiversity of flora, fauna, and beneficial microorganisms [44].

The most common trees found in Ecuadorian coffee plantations are a mix of fruit trees (*Musa paradisiaca, Carica papaya, Citrus limon, Psidium guajava, Inga edulis*, and *Theobroma cacao*) and timber trees (*Cordia alliodora* and *Ochroma* sp) [44]. In Galapagos is common to find other endemic species such as *Scalesia pedunculata, Psidium galapageium,* and *Zanthoxilum fagara* [45]. **Figure 2** gives an idea of the biodiversity around coffee trees.

Because of all positive factors, the agroforestry system is more sustainable. However, it is important that farmers adapt this system to their specific conditions to avoid competition between species for nutrients and water. This competition could decrease production. The tree density recommended by the sustainable agriculture network (SAN) is 40%. Higher densities subtract sunlight from coffee, producing opposite effects [46]. The choice of shade trees is also important. Trees with deep and widely branched roots are generally preferred. Leguminous trees are also relevant for their ability to fix nitrogen. In Ecuador, the leguminous Guaba tree (*Inga edulis*) is widely found also because its fruit is locally consumed [47].

Coffee trees tolerate a wide range of soils if they are deep, porous, well-drained, and well balanced for their texture. Coffee is not very demanding in soil fertility, and it can be cultivated in fertile as well as in poor soils even in acidic soils. Ecuadorian volcanic soils are particularly well suited for coffee [48]. Nonetheless, the production of green coffee leads to the depletion of nutrients. This depletion needs to be compensated by appropriate fertilization to keep a constant and high production. Proper nutrition is important for vigorous plants. Parameters such as the age of the coffee trees, the planting density, and the degree of intensification must be considered [46]. It is advisable to take soil samples before applying fertilizers. Foliar fertilization is often used to compensate deficiencies in micronutrients like zinc, boron, iron, and manganese [49].

Figure 2.
Coffee trees under agroforestry system. Intag-Ecuador. (Source: Sania Ortega).

Ecuadorian rural farms do not have nearby laboratories to monitor the nutrient content in the soil. So, fertilization and fumigation are based on farmers' intuition and experience. Unfortunately, this is a problem that could seriously affect the quality of the soils as well as coffee production. Another problem is the lack of records of the treatments applied.

Although correct fertilization can supply any nutrient deficiency, to guarantee sustainability it is important to preserve the soil microbiota. Microorganisms play a very important role in soil fertility and crop production because of their ability to promote plant growth, enhance biotic and abiotic stress resistance, and facilitate and improve the absorption of nutrients by the root [50].

Plant growth-promoting microorganisms, and arbuscular mycorrhizal fungi (AMF) have been used in coffee trees to improve productivity and reduce the application of chemical fertilizers [50, 51]. In a study carried out in Mexico, for example, it was shown that the inoculation of coffee seedlings with *Azospirillum* sp., *Glomus intraradices,* and *Azotobacter* sp. improved root structure [52]. *Azospirillum* promoted the formation of root hairs, which in turn facilitated the plant-mycorrhizal association. This association increased the uptake of phosphorus and the secretion of radical exudates that favored the development of *Azotobacter* known for its ability to fix nitrogen. In conclusion, the colonization of the roots by beneficial microorganisms stimulates plant growth and improves nutrients uptake obtaining healthy plants. The presence of beneficial microorganisms in the rhizosphere decreases the presence of other non-beneficial or plant pathogens. Microorganisms produce elicitors such as volatile organic compounds, antimicrobials, and/or competition. These elicitors can induce the expression of pathogenesis-related genes in plants through induced systemic resistance or acquired systemic resistance channels [53].

A critical task for coffee growers around the world is the control of pathogens and diseases. The biotrophic fungus *Hemileia vastatrix*, or coffee leaf rust is considered one of the most devastating in Latin America [54]. In Ecuador, in 2013, it caused losses greater than 50% of total production [55]. To save production and reduce the incidence of rust, in 2012–2021 Ecuadorian coffee-cocoa reactivation program, coffee growers were encouraged to plant species with resistant genotypes as a strategy to control coffee rust [26]. However, some coffee farmers fear lowering cup quality by replacing one variety with another. Among the agricultural practices used to stop coffee rust, the use of shade trees for temperature control is the most common. However, this management practice is effective only if it does not produce excess humidity or decrease photosynthesis, which promotes the growth of the fungus [56]. Another common practice to reduce fungus colonization is the application of silicates at the foliar level. Chemical control with cupric fungicides is also widely used despite the increasing interest in organic certification [57]. The biological control of coffee leaf rust by antagonistic bacteria was also studied. Bacteria belonging to the genus *Bacillus* and *Pseudomonas* showed a great potential for rust control in Brazil [58]. Also, the entomopathogenic and mycoparasite fungus *Lecanicillium lecanii* proved to be effective in rust control [59].

On the other hand, coffee berry borer (*Hipothenemus hampei*) is the most challenging insect pest of coffee throughout the globe. Adult females bore the berry and deposit eggs inside it, altering the organoleptic properties of the bean and reducing the selling price between 20% and 40% [60]. In addition, it causes premature fall of youngberries and the increased susceptibility of infested ripe berries to fungus or bacterial infection [61]. Due to its nature and behavior, the borer is very difficult to eradicate. The fact that it lives inside the fruit makes contact insecticides ineffective. Organochlorine and organophosphate insecticides are the most widely used but produce high environmental costs [62]. In countries like Brazil, Mexico, and Colombia traps have been designed. The most technological ones allow to capture up to 10,000 adults per day [61]. Biological control with entomopathogenic fungi has also been an option, although it is not sufficiently effective until now. The most used microorganism to control insect pests is *Beauveria bassiana* [63].

With this overview, obtaining quality coffee is not an easy task and requires extensive technical and scientific knowledge. The error of the programs carried out by the Ministry of Agriculture was to generalize production and not consider the unique characteristics of each ecosystem. Successful models in other countries are not always adapted to the reality of each region. In conclusion, there are research lines that must be established to select varieties with superior characteristics, understand the ecological relationship of coffee in each ecosystem and isolate native microorganisms useful to improve plant growth and pest. It is also important to register the treatments and practices carried out by the farmers and learn from their experiences.

5. Technical and scientific aspects of coffee processing and wastes disposal

The cup of coffee has a process behind it that varies according to the region, the variety of coffee planted, and the use of the different post-harvest treatments. All these factors give each coffee a unique flavor.

Coffee trees, depending on the specific variety, take between 3 and 4 years to bear the first fruits. Generally, coffee cherry ripening is faster in lower and warmer areas. Nonetheless, slow ripening is more advisable to achieve better organoleptic characteristics [64]. To guarantee coffee quality, it is important to harvest only ripe fruits. The coffee tree declines its productivity after 20 years [65]. Therefore, it is important to renew coffee plantations.

The process of harvesting coffee beans can be carried out by different methods, among which are manual or mechanical. In Ecuador, a manual process is the most used. Workers collect the coffee berry, avoiding collecting green ones and discarding the grains that are dry or damaged. All coffee fruits are collected in plastic handcrafted containers and transported to the classification area. Harvesting depends on the labor and skill acquired to select the best fruits. In general, it is hard work since the collectors must walk long distances on slopes where a big part of coffee plantations are located.

After fruits collection, the flotation technique is used for the selection of the beans. It consists in covering the cherries with water. Contaminants such as stones, garbage, and floats are discarded. Subsequently, a second review of the fruits is carried out, spreading them on African beds or similar structures to discard those that are not cherry-colored. All discarded fruits (pasilla) are considered inferior in quality and therefore have a lower price in the market.

Post-harvest treatments are part of the coffee production process, the latter can be classified into three different types of processing: wet or washed, dry or natural, and semi-dry or honey. The country's coffee growers carry out empirical experiments to determine which treatment provides the best quality results. In general, dry processes are applied in the Robusta variety, and wet ones in Arabica [66, 67]. The different steps of each process are summarized in **Figure 3**.

All treatments share the harvest and flotation stage. The natural or dry process is considered the simplest and most traditional at the national level since it consists of the direct drying of the fruits and the subsequent removal of the dry pericarp by manual or mechanical action [68]. On the other hand, the semi-dry process has a previous stage of mechanical pulping to remove the pericarp before drying [67]. Finally, the wet process requires a fermentation step which needs robust control. The quality of the coffee obtained from the wet process is generally higher compared to the other processes [68].

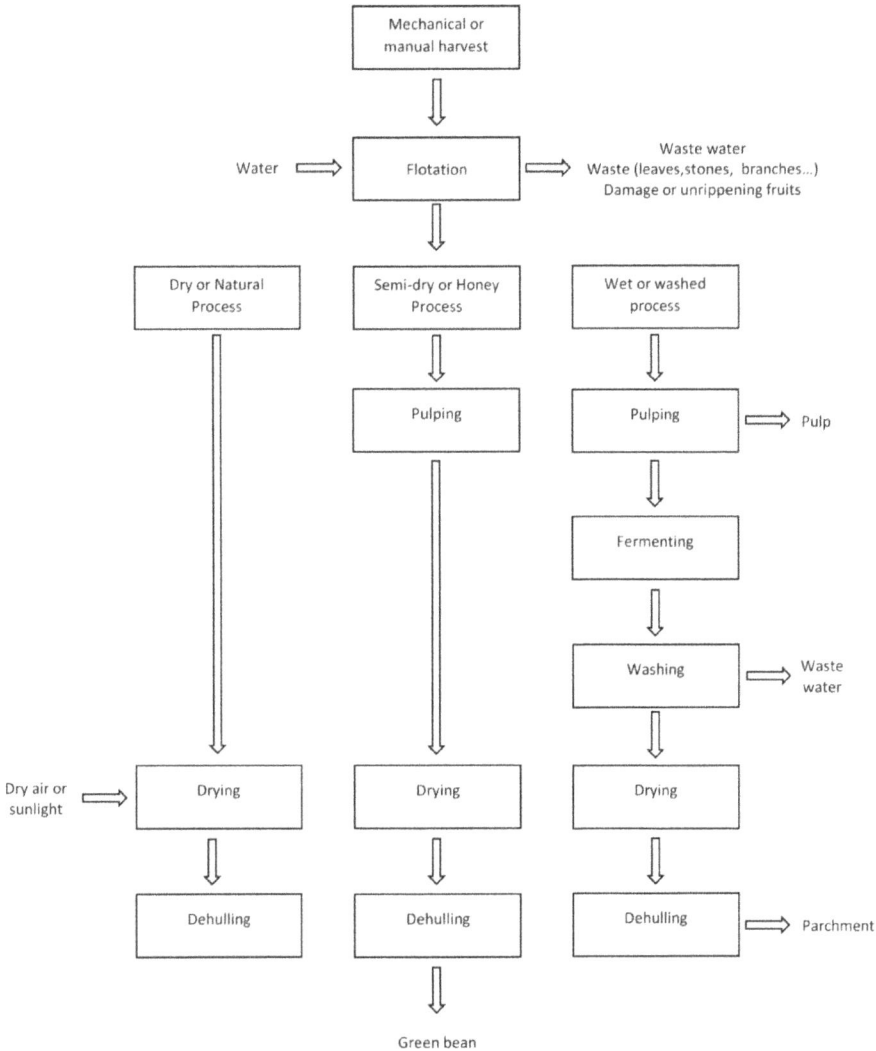

Figure 3.
Coffee processing technologies.

5.1 Dry or natural process

Natural processing is considered the oldest and most traditional technology. It is relatively simple and inexpensive. Previously selected cherries went through a drying process, to finally be shelled and pulped. The drying process prevents the growth of microorganisms. It is done under the sun or through air dryers that allow reaching a humidity of 10–12%, which is considered a standard measure for the coffee to retain its volatile compounds until roasting [67, 68].

In Ecuador, the Robusta coffee is dried directly under the sunlight. The advantage of this process is the cost. However, there are several disadvantages such as the time of the process which depends on the weather conditions, and the constant control required to prevent damage due to dust, rain, or storms. This process ends with the extraction of the pericarp and the dry pulp to obtain only the green coffee beans that will be later stored. This can be done manually through a mortar or threshing machine, or through mechanical hullers.

13

5.2 Wet or washed process

The wet process is the treatment with the best results in terms of coffee quality [69]. The main difference with the dry process is the pulped step. The pulping is carried out mechanically and consists of the removal of the exocarp and part of the mesocarp of the coffee cherry. **Figures 4** and **5**, show a classical pulped machine used by smallholders. The pulp is squeezed through a rotating disc or drum. This process must be carried out in a way that does not damage the bean which could lead to microbial attack or contamination. The part of the remaining mesocarp is the mucilage, which will be important in the subsequent fermentation process [67, 70].

Mucilage fermentation is the important turning point for wet production due to the quality indicators that it provides such as aroma and flavor [70, 71]. The fermentation process is important to degrade the mucilage resulting from the previous process, containing a large amount of pectin, starch, and cellulose; being an ideal substrate for yeasts and bacteria [71]. In Ecuador, a metagenomic study on the fermentation processes of coffee was developed confirming the presence of

Figure 4.
Pulped Machine in Piedra Grande San Jerónimo's Farm, Lita-Imbabura (source: Sabastian Obando).

Figure 5.
Pulped process of a family agroforestry system. Intag-Ecuador. (source: Sania Ortega).

enterobacteria, lactic acid bacteria, and yeasts as the majority of microorganisms' groups [72]. Microorganism isolation and evaluation of fermentation ability must be the next step to improve flavor quality.

Microorganisms are the secret of fermentation. Its metabolic routes produce secondary metabolites and volatile compounds associated with aromas and flavors [73].

Studies have been carried out to isolate different microorganisms and evaluate their ability to produce volatile compounds and how they affect coffee quality [74, 75]. Alcohols are abundant volatile metabolites that fulfill various functions such as providing fruity aromas and flavors, contributing to physical characteristics *e.g.,* adding viscosity to the beverage, and even serving as a solvent for other volatile compounds. However, some compounds such as aldehydes are variable and can give desired characteristics e.g., fruity, and almond aromas. But also, undesirable characteristics like the pungent taste. Finally, it is responsible for providing bitterness, an important characteristic in the evaluation of coffee cups [76].

The washing process must be carried out immediately after the fermentation process is completed to avoid an overfermentation and production of propionic and butyric acid that are related to onion flavors and aromas. Washing is done with drinking or irrigation water seeking to remove all the mucilage resulting from the fermentation process before drying. The drying and shelling are carried out under the same conditions as the dry or natural process [67, 68]. **Figure 6** is a classical air-drying installation.

5.3 Semi-dry or honey process

The semi-dry process is a combination of the two previous processes. The bean with the layer of mucilage is left to dry directly in the sun, as in both processes. This step allows the layer of mucilage to impregnate the bean giving it a color and texture equal to honey. It is a process that requires less control. The fermentation

Figure 6.
Sun-drying coffee beans. (Source Sebastian Obando).

process is considered to occur in the drying stage. Microbial inoculants have been also studied in this process [77, 78].

This process is the most recommended in Ecuador since it allows to obtain high quality and produces less wastewater. Some types of coffee processed with the wet and honey method have been awarded with the golden cup [42].

After coffee bean processing, bean coffee could be roasted or stored. Storage conditions must be controlled to prevent fungal growth and mycotoxin production [79].

Like cultivation, coffee processing requires extensive knowledge and special attention since it directly influences the quality of the final product. Each producer maintains in reserve the details of the processing, especially in the fermentation stage. However, cooperation between producers is a key factor in marketing. Coffee Associations often fail to meet the demand of international markets due to differences in quality obtained between partners.

As in cultivation, coffee producers experiment by varying production methods or parameters during the process. Although the experimentation carried out is a positive aspect, it is necessary to standardize the production to guarantee the same quality in all harvests. Research is also an essential component of development.

5.4 Waste disposal

Each process generates a different type and volume of waste such as water, pulp, and parchment (**Figure 1**). It is estimated that ¾ of the volume of the total beans harvested are residues [80]. Since in Ecuador the cultivation of coffee is a family and rural activity, the use of sophisticated technologies for the valorization of residues is not applicable. The most suitable technologies are related to reuse in agriculture, animal feed, and energy production [81].

A large part of the farmers processes their own coffee on the farm, so the waste is managed internally. A common practice is to spread the residues directly on the fields and let them naturally degrade. However, the application of not completely degraded residues produces adverse effects, including phytotoxicity [82]. On the other hand, this practice facilitates the spread of pests from infected and discarded fruits, and bad odors, among others. In addition, this bad practice produces contamination of water sources. Most of the farms are located on hills and when it rains the water carries pollutants to the lower areas that end up polluting rivers.

Other coffee growers take the harvested beans to the collection centers to be processed afterward. These collection centers usually have a place for composting. Compost obtained is then sold as an amendment to use in coffee or other minor crops. Data on the quality of this amendment is not available. Generally, composting process is carried out under partial or not controlled conditions. Something important to highlight is that each farm must be concerned about its sustainability to be competitive. Organic certifications are always more important to gain and guarantee a marketplace. On the other hand, chemical fertilizers are expensive. Therefore, taking advantage of nutrient-rich waste is the most viable option for self-sustainability.

Coffee by-products are characterized by a high concentration of nutrients and other compounds such as polyphenols and caffeine that in high concentrations could be phytotoxic [80]. For this reason, it is important to a stabilization treatment before its use in agriculture. Composting remains the best and cheapest way to achieve this aim. However, it must be a controlled process to guarantee phytotoxic reduction and pathogens elimination.

Because composting is a microbial degradation process. The selection of specific degrading microorganisms could be a good option to improve compost quality and reduce composting time. Compost can also be used as a strategy to introduce PGPR and biocontrol agents. A similar experience was achieved by the Italian olive industry which residues are very similar in composition to those of coffee [83].

Generally, coffee growers mix coffee by-products with other agricultural residues from minor crops and manure from raising pigs, chickens, guinea pigs, and cows. These processes are not technical and therefore the results obtained may be variable or not satisfactory. For this reason, boosting the correct management of composting technology would require appropriate training for groups. This should include appropriate training on quality control of the final product to guarantee the reduction of phytotoxicity, the elimination of pathogens, and the stabilization of the compost. Good results of compost application in rural areas were obtained in Vietnam [84].

Another re-utilization process is to use by-products as animal feed. The presence of tannins and caffeine diminishes the acceptability and palatability of husk by animals. So, the degradation of caffeine by microorganisms, especially bacteria, needs further studies.

In conclusion, composting is the most applicable technology to coffee waste management because it only requires a space that farms generally have and common work tools to remove de compost. It also helps to return to the soil part of the nutrients extracted by agriculture. Other technologies are too expensive and require big quantities to recover the investment. As in cultivation and processing, waste disposal should also be linked to research programs that over time can provide alternative solutions contributing to rural and sustainable development.

6. Conclusions and policy implications

Coffee is a strategic crop; it has a growing world market and therefore great economic potential. It also has environmental benefits that distinguish it from other expansive crops such as palm and banana.

On the other hand, Ecuador has optimal geographical and climatic conditions for growing coffee but has a lower production (3–5 quintals/ha) compared to other producing countries in the region such as Brazil and Colombia (35–40 quintals/ha) [20]. These competitive disadvantages prevent Ecuador from covering the market demand, which has affected the coffee trade [17]. The low productivity can be explained by many factors such as poverty in rural areas, lack of trained workers, inadequate management of pests and diseases, the presence of aged coffee plantations, insufficient infrastructure technology for post-harvest processes, and the lack of effective marketing channels [15]. To overcome these deficiencies, it is essential to improve the quality of life of farmers by guaranteeing access to basic services and education. Farmer's income must be protected with adequate economic policies. This will allow new generations to see agriculture as a profitable livelihood and assure sustainability.

Scientific research is also important to overcome problems like pest control and productivity. Reactivation programs must include the active participation of research centers and not just incentives and subsidies.

The sustainable development of coffee farming in rural areas does not necessarily require large investments, but it does require cooperation between farmers and research centers to guarantee knowledge transfer. The variability of coffee quality between farmers fails to meet market demand. Coffee producers tend to compete rather work as a team and help others to achieve quality, so cooperation is a point of force.

Under current conditions, Ecuador is not competitive in terms of volume due to a lack of technology and workforce, however, it can be very competitive in terms of quality thanks to the variability of ecosystems that give coffee special characteristics.

The results of the coffee-cocoa reactivation program established by the Ministry of Agriculture of Ecuador in the years 2012–2021 are expected to show appreciable improvements in productivity. Furthermore, this program is expected to allow the renovation of coffee plantations, the technical training of farmers, and the implementation of modern infrastructure.

Making changes in public policies to comply with territorial development programs that are based on sustainable development objectives is needed. Public policy in the "Socio Bosque" Program, for example, should be strengthened to generate incentives for farmers who have or opt for agroecological plantations and have organic certifications. In this way, the program would ensure the maintenance of primary forests while increasing economic income for coffee-growing families. In this way, conservation would be guaranteed along with the improvement of life quality.

Acknowledgements

We want to thank the associations of organic coffee producers Bosque Nublado Golondrinas and Río Intag for opening the doors of their farms and sharing with us their day-by-day efforts in this wonderful activity.

Author details

Echeverría María Cristina[1*], Ortega-Andrade Sania[1], Obando Sebastián[1] and Marco Nuti[2]

1 Technical University of the North, Ibarra, Ecuador

2 Sant'Anna School of Advanced Studies, Pisa, Italy

*Address all correspondence to: mecheverria@utn.edu.ec

IntechOpen

References

[1] Nieber K. The impact of coffee on health. Planta Medica. 2017;**83**:1256-1263. DOI: 10.1055/s-0043-115007

[2] Crop year production by country. International Coffee Organization [Internet]. 2021. Available from: https://www.ico.org/prices/po-production.pdf [Accessed: January 26, 2021]

[3] World Coffee Consumption. International Coffee Organization [Internet]. 2021. Available from: https://www.ico.org/prices/new-consumption-table.pdf [Accessed: January 26, 2021]

[4] Five-Year-Action Plan for the International Coffee Organization [internet]. 2017 Available from: https://www.ico.org/documents/cy2016-17/wp-council-280-r1e-five-year-action-plan.pdf [Accessed: January 26, 2021]

[5] Informe de rendimientos objetivos de café (grano oro). Ministerio de Agricultura y Ganadería. Sistema de Información Pública Agropecuaria [Internet]. 2019. Available from: http://sipa.agricultura.gob.ec/descargas/estudios/rendimientos/cafe/rendimiento_cafe_2019.pdf [Accessed: January 26, 2021]

[6] Exports of all forms of coffee by exporting countries to all destinations. International Coffee Organization [Internet]. 2021. Available from: https://www.ico.org/prices/m1-exports.pdf [Accessed: January 26, 2021]

[7] Alemu MM. Effect of tree shade on coffee crop production. Journal of Sustainable Development. 2015;**8**(9):66. DOI: 10.5539/jsd.v8n9p66

[8] Albrecht A, Kandji ST. Carbon sequestration in tropical agroforestry systems. Agriculture, Ecosystems & Environment. 2003;**99**(1-3):15-27. DOI: 10.1016/S0167-8809(03)00138-5

[9] Torres Castillo NE, Melchor Martínez EM, Ochoa Sierra JS, Ramirez-Mendoza RA, Parra Saldívar R, Iqba HMN. Impact of climate change and early development of coffee rust: An overview of control strategies to preserve organic cultivars in Mexico. Science of the Total Environment. 2020;**738**:140-225. DOI: 10.1016/j.scitotenv.2020.140225

[10] Jaramillo J, Chabi-Olaye A, Kamonjo C, Jaramillo A, Vega FE, Poehling HM, et al. Thermal tolerance of the coffee Berry Borer Hypothenemus hampei: Predictions of climate change impact on a tropical insect pest. Plos One. 2009;**4**(8):64-87. DOI: 10.1371/journal.pone.0006487

[11] Silva MDC, Várzea V, Guerra-Guimarães L, Azinheira HG, Fernandez G, Petitot AS, et al. Coffee resistance to the main diseases: Leaf rust and coffee berry disease. Brazilian Journal of Plant Physiology. 2006;**18**(1):119-147. DOI: 10.1590/S1677-04202006000100010

[12] Davis AP, Gole TW, Baena S, Moat J. The impact of climate change on indigenous arabica coffee (*Coffea arabica*): Predicting future trends and identifying priorities. Plos One. 2012;7(11):479-481. DOI: 10.1371/journal.pone.0047981

[13] Rao MV, Rice RA, Fleischer RC, Muletz-Wolz CR. Soil fungal communities differ between shaded and sun-intensive coffee plantations in El Salvador. Plos One. 2020;**15**(4):1-19. DOI: 10.1371/journal.pone.0231875

[14] Zijlstra C, Lund I, Justesen AF, Nicolaisen M, Jensen PK, Bianciotto V, et al. Combining novel monitoring tools and precision application technologies for integrated high-tech crop protection in the future (a discussion document).

Pest Management Science. 2011;**67**(6): 616-625. DOI: 10.1002/ps.2134

[15] Mero Loor KA, Muñoz González R. Economía política del desarrollo: claves del sector cafetalero para el desarrollo territorial de Manabí, Ecuador. Revista Observatorio de la Economía Latinoamericana. 2018;**1**:1-11

[16] Cifras Agroproductivas. Sistema de Información Pública Agropecuaria [Internet]. 2020. Available from: http://sipa.agricultura.gob.ec/index.php/cifras-agroproductivas [Accessed: January 26, 2021]

[17] Exportaciones. Asociación Nacional Ecuatoriada de Café [Internet]. 2021. Available from: https://www.anecafe.org.ec/estadisticas/ [Accessed: January 26, 2021]

[18] Indicadores de pobreza y desigualdad. Instituto Nacional de Estadística y Censos [Internet]. 2021. Available from: https://www.ecuadorencifras.gob.ec/documentos/web-inec/POBREZA/2021/Diciembre-2021/202112_PobrezayDesigualdad.pdf [Accessed: January 26, 2021]

[19] La política agropecuaria Ecuatoriana: Hacia el desarrollo territorial rural sostenible 2015-2025. Ministerio de Agricultura, Ganadería y Pesca [Internet]. 2014. Available from: http://www2.competencias.gob.ec/wp-content/uploads/2021/03/03-01PPP2016-POLITICA02.pdf [Accessed: January 26, 2021]

[20] Ecuador en una mirada. Food and Agriculture Organization [Internet]. 2022. Available from: https://www.fao.org/ecuador/fao-en-ecuador/ecuador-en-una-mirada/es/ [Accessed: January 26, 2021]

[21] Mittermeier RA, Turner WR, Larsen FW, Brooks TM, Gascon C. Global biodiversity conservation: The

critical role of hotspots. In: Zachos FE, Jan Christian H, editors. Biodiversity Hotspots. Berlin, Heidelberg: Springer; 2011. pp. 3-22

[22] Programa de protección de bosques, Socio Bosque. Ministerio del Ambiente [Internet]. 2022. Available from: http://sociobosque.ambiente.gob.ec/node/173 [Accessed: January 26, 2021]

[23] Plan Nacional de Restauración Forestal. Ministerio del Ambiente [Internet]. 2017. Available from: http://sociobosque.ambiente.gob.ec/files/images/articulos/archivos/amrPlanRF.pdf [Accessed: January 26, 2021]

[24] Viteri Salazar O, Ramos-Martín J. Organizational structure and commercialization of coffee and cocoa in the northern Amazon Region of Ecuador. Revista Nera. 2017;**35**:266-287

[25] Turner EC, Snaddon JL, Ewers RM, Fayle TM, Foster WA. The impact of oil palm expansion on environmental change: Putting conservation research in context. In: Dos Santos Bernardes MA, editor. Environmental Impact of Biofuels. London: IntechOpen; 2011. pp. 202-263. DOI: 10.5772/20263

[26] Chain-Guadarrama A, Martínez-Salinas A, Aristizábal N, Ricketts TH. Ecosystem services by birds and bees to coffee in a changing climate: A review of coffee berry borer control and pollination. Agriculture. Ecosystems & Environment. 2019;**280**:53-67. DOI: 10.1016/j.agee.2019.04.011

[27] Café-Cacao. Ministerio de Agricultura [internet]. 2022. Available from: https://www.agricultura.gob.ec/cafe-cacao/ [Accessed: January 26, 2021]

[28] Ortega Andrade S, Páez GT, Feria TP, Muñoz J. Climate change and the risk of spread the fungus from the high mortality of Theobroma cacao in

Latin América. Neotropical Biodiversity. 2017;**3**(1):30-40. DOI: 10.1080/23766808.2016.1266072

[29] Venegas Sánchez S, Orellana Bueno D, Pérez JP. La realidad Ecuatoriana en la producción de café. Revista Científica Mundo de la Investigación y el Conocimiento. 2018;**2**:72-91

[30] Borrella I, Mataix C, Carrasco-Gallego R. Smallholder farmers in the speciality coffee industry: Opportunities, constraints and the businesses that are making it possible. IDS Bulletin. 2015;**46**(3):29-44

[31] Hellin J, Higman S. Smallholders and niche markets: Lessons from the Andes. Agricultural Research & Extension Network. 2002;**118**:1-16

[32] Poltronieri P, Rossi F. Challenges in specialty coffee processing and quality assurance. Challenges. 2016;7:1-22. DOI: 10.3390/challe7020019

[33] Ecuador Galapagos Coffee Production and Commercialization. USDA Foreing Agricultural Service [Internet]. 2013. Available from: https://apps.fas.usda.gov/newgainapi/api/report/downloadreportbyfilename?filename=Ecuador%20Galapagos%20Coffee%20Production%20and%20Commercialization%20%20_Quito_Ecuador_1-25-2013.pdf [Accessed: January 26, 2021]

[34] Superficie, producción y ventas según región y provincia Café (grano oro). Instituto Nacional de Estadísticas y Censos (INEC) [Internet]. 2021. Available from: https://www.ecuadorencifras.gob.ec/estadisticas-agropecuarias-2/ [Accessed: January 26, 2021]

[35] Ludwig IA, Mena P, Calani L, Cid C, del Rio D, Lean ME, et al. Variations in caffeine and chlorogenic acid contents of coffees: What are we drinking? Food

and Function. Royal Society of Chemistry. 2014;**5**:1718-1726. DOI: 10.1039/C4FO00290C

[36] Duchicela Guambi LA, Farfán Talledo DS, García Ávila EL. Calidad organoléptica del café (*Coffea arabica* L.) en las zonas centro y sur de la provincia de Manabí, Ecuador. Revista española de estudios agrosociales y pesqueros. 2016;**244**:15-34

[37] Bote AD, Struik PC. Effects of shade on growth, production and quality of coffee (*Coffea arabica*) in Ethiopia. Journal of Horticulture and Forestry. 2011;**3**:336-341

[38] Pham Y, Reardon-Smith K, Mushtaq S, Cockfield G. The impact of climate change and variability on coffee production: A systematic review. Climatic Change. 2019;**156**:609-630. DOI: 10.1007/s10584-019-02538-y

[39] Duicela-Gambi L A, Velásquez S. Organoleptic quality of arabian coffees in reation to the varieties and altitudes of the growing areas, Ecuador [Internet]. 2017. Available from: https://www.researchgate.net/publication/321732953. [Accessed: January 26, 2021]

[40] Guía de variedades de Café. Asociación Nacional del Café (ANACAFE) [internet]. 2020. Available from: https://www.anacafe.org/uploads/file/8ee92f426ab648318001477e70d0bbe1/Gu%c3%ada-de-variedades-Anacaf%c3%a9-2020.pdf [Accessed: January 26, 2021]

[41] Leroy T, Ribeyre F, Bertrand B, Charmetant P, Dufour M, Montagnon C, et al. Genetics of coffee quality. Brazilian Journal of Plant Physiology. 2006;**18**(1):229-242. DOI: 10.1590/S1677-04202006000100016

[42] Concurso Taza Dorada. ANECAFE [Internet]. 2021. Available from: https://www.anecafe.org.ec/taza-dorada/ [Accessed: January 26, 2021]

[43] Duicela Guambi LA, Andrade Moreano J, Farfán Talledo DS. Calidad organoléptica, métodos de beneficio y cultivares de café robusta (Coffea canephora Pierre ex Froehner) en la amazonía del Ecuador. Revista Iberoamericana de Tecnología Postcosecha. 2018;**19**(2):1-17

[44] de Beenhouwer M, Aerts R, Honnay O. A global meta-analysis of the biodiversity and ecosystem service benefits of coffee and cacao agroforestry. Agriculture, Ecosystems and Environment. 2013;**175**:1-7. DOI: 10.1016/j.agee.2013.05.003

[45] Infante B M, Caracterización de la composición botánica dentro de una plantación modelo de café bajo un sistema agroforestal en la zona de El Camote, Isla Santa Cruz, Galápagos [thesis]. Quito: Universidad Central del Ecuador; 2019

[46] Lambot C, Herrera JC, Bertrand B, Sadeghian S, Benavides P, Gaitán A. Cultivating Coffee Quality-Terroir and Agro-Ecosystem. In: Folmer B, editor. The Craft and Science of Coffee. 1st ed. Cambridge-Massachusetts: Academic Press; 2017. pp. 17-49

[47] Vargas-Tierras YB, Prado-Beltrán JK, Nicolalde-Cruz JR, Casanoves F, de Melo V-FE, Viera-Arroyo WF. Characterization and role of Amazonian fruit crops in family farms in the provinces of Sucumbíos and Orellana (Ecuador). Ciencia y Tecnología Agropecuaria. 2018;**19**:501-515

[48] Espinosa J, Moreno J. The use of land. In: Espinosa J, Moreno J, Bernal G, editors. The Soils of Ecuador. 1st ed. Luxemburg: Springer; 2018. pp. 151-162

[49] Clemente JM, Martinez HEP, Pedrosa AW, Poltronieri Neves Y, Cecon PR, Jifon JL. Boron, copper, and zinc affect the productivity, cup quality, and chemical compounds in coffee beans. Journal of Food Quality. 2018:1-14

[50] Urgiles-Gómez N, Avila-Salem ME, Loján P, Encalada M, Hurtado L, Araujo S, et al. Plant growth-promoting microorganisms in coffee production: From isolation to field application. Agronomy. 2021;**11**(8):1-12. DOI: 10.3390/agronomy11081531

[51] Hernández-Acosta E, Trejo-Aguilar D, Rivera-Fernández A, Ferrera-Cerrato R. Arbuscular mycorrhiza as a biofertilizer in production of coffee. Terra Latinoamericana. 2020;**38**(3): 613-628

[52] Adriano Anaya MDL, Gálvez RJ, Hernández Ramos C, Figueroa MS, Monreal Vargas CT. Biofertilizer of organic coffee in stage of seedlings in Chiapas, México. Revista mexicana de ciencias agrícolas. 2011;**2**(3):417-431

[53] Enebe MC, Babalola O. The impact of microbes in the orchestration of plants' resistance to biotic stress: A disease management approach. Applied Microbiology and Biotechnology. 2019;**103**(1):9-25. DOI: 10.1007/s00253-018-9433-3

[54] Rhiney K, Guido Z, Knudson C, Avelino J, Bacon CM, Leclerc G, et al. Epidemics and the future of coffee production. Proceedings of the National Academy of Sciences. 2021;**118**(27):1-10. DOI: 10.1073/pnas.2023212118

[55] MAGAP declara emergencia para controlar roya del café. Ministerio de Agricultura [Internet]. 2013. Available from: https://www.agricultura.gob.ec/magap-decreto-estado-de-emergencia-para-controlar-la-roya-del-cafe/ [Accessed: January 28, 2021]

[56] Avelino J, Zelaya H, Merlo A, Pineda A, Ordóñez M, Savary S. The intensity of a coffee rust epidemic is dependent on production situations. Ecological Modelling. 2006;**197**(3-4): 431-447. DOI: 10.1016/j.ecolmodel.2006.03.013

[57] Pereira DR, Nadaleti DH, Rodrigues EC, da Silva AD, Malta MR, de Carvalho SP, et al. Genetic and chemical control of coffee rust (Hemileia vastatrix B erk et B r.): Impacts on coffee (Coffea arabica L.) quality. Journal of the Science of Food and Agriculture. 2021;**101**(7):2836-2845. DOI: 10.1002/jsfa.10914

[58] de Resende ML, Pozza EA, Reichel T, Botelho D. Strategies for coffee leaf rust management in organic crop systems. Agronomy. 2021;**11**(9):1865. DOI: 10.3390/agronomy11091865

[59] Zewdie B, Tack AJ, Ayalew B, Adugna G, Nemomissa S, Hylander K. Temporal dynamics and biocontrol potential of a hyperparasite on coffee leaf rust across a landscape in Arabica coffee's native range. Agriculture, Ecosystems & Environment. 2021;**311**:107297

[60] Abate B. Coffee Berry Borer, Hypothenemus hampei(Ferrari) (Coleoptera: Scolytidae): A challenging coffee productions and future prospects. American Journal of Entomology. 2021;**5**(3):39-46

[61] Lemma DT, Abewoy D. Review on integrated pest management of coffee Berry Disease and Coffee Berry Borer. International Journal of Plant Breeding and Crop Science. 2021;**8**(1):1001-1008

[62] Júnior SDMD, Soares WS, Celoto FJCJ, Fernandes FL, Oliveira MMF, Botrel GBB. Resistance and effect of insecticide-treated coffee berries of different varieties to the penetration of Hypothenemus hampei (Coleoptera: Curculionidae: Scolytinae). Coffee Science. 2021;**16**:161874

[63] Posada-Flórez FJ. Production of Beauveria bassiana fungal spores on rice to control the coffee berry borer, Hypothenemus hampei, in Colombia. Journal of Insect Science. 2008;**8**(1):41. DOI: 10.1673/031.008.4101

[64] Geromel C, Ferreira LP, Davrieux F, Guyot B, Ribeyre F, dos Santos SMB, et al. Effects of shade on the development and sugar metabolism of coffee (*Coffea arabica* L.) fruits. Plant Physiology and Biochemistry. 2008;**46**(5-6):569-579. DOI: 10.1016/j.plaphy.2008.02.006

[65] Kudama G. Factors influencing coffee productivity in Jimma zone, Ethiopia. World Journal of Agricultural Sciences. 2019;**15**(4):228-234. DOI: 10.5829/idosi.wjas.2019.228.234

[66] Vincent JC. Green coffee processing. In: Clarke RJ, Macrae R, editors. Coffee Technology. Dordrecht: Springer; 1987. pp. 1-33. DOI: 10.1007/978-94-009-3417-7_1

[67] Ghosh P, Venkatachalapathy N. Processing and drying of coffee–a review. International Journal of Engineering Research & Technology. 2014;**3**(12):784-794

[68] de Melo PGV, de Carvalho NDP, Júnior AIM, Vásquez ZS, Medeiros AB, Vandenberghe LP, et al. Exploring the impacts of postharvest processing on the aroma formation of coffee beans–A review. Food Chemistry. 2019;**272**:441-452. DOI: 10.1016/j.foodchem.2018.08.061

[69] Subedi RN. Comparative analysis of dry and wet processing of coffee with respect to quality and cost in Kavre District, Nepal: A case of Panchkhal Village. International Research Journal of Applied and Basic Sciences. 2011;**2**(5):181-193

[70] Lee LW, Cheong MW, Curran P, Yu B, Liu SQ. Coffee fermentation and flavor–An intricate and delicate relationship. Food Chemistry. 2015;**185**:182-191. DOI: 10.1016/j.foodchem.2015.03.124

[71] Avallone S, Guyot B, Brillouet JM, Olguin E, Guiraud JP. Microbiological and biochemical study of coffee fermentation. Current Microbiology. 2001;**42**(4):252-256. DOI: 10.1007/s002840110213

[72] Pothakos V, De Vuyst L, Zhang SJ, De Bruyn F, Verce M, Torres J, et al. Temporal shotgun metagenomics of an Ecuadorian coffee fermentation process highlights the predominance of lactic acid bacteria. Current Research in Biotechnology. 2020;**2**:1-15. DOI: 10.1016/j.crbiot.2020.02.001

[73] Haile M, Kang WH. The role of microbes in coffee fermentation and their impact on coffee quality. Journal of Food Quality. 2019;**2019**:1-6. DOI: 10.1155/2019/4836709

[74] Bressani APP, Martinez SJ, Evangelista SR, Dias DR, Schwan RF. Characteristics of fermented coffee inoculated with yeast starter cultures using different inoculation methods. LWT-Food Sciences and Technology. 2018;**92**:212-219. DOI: 10.1016/j.lwt.2018.02.029

[75] Martinez SJ, Bressani APP, Dias DR, Simão JBP, Schwan RF. Effect of bacterial and yeast starters on the formation of volatile and organic acid compounds in coffee beans and selection of flavors markers precursors during wet fermentation. Frontiers in Microbiology. 2019;**10**:1287. DOI: 10.3389/fmicb.2019.01287

[76] da Silva VA, de Melo PGV, de Carvalho NDP, Rodrigues C, Pagnoncelli MGB, Soccol CR. Effect of co-inoculation with Pichia fermentans and Pediococcus acidilactici on metabolite produced during fermentation and volatile composition of coffee beans. Fermentation. 2019;**5**(3):67. DOI: 10.3390/fermentation5030067

[77] Evangelista SR, da Cruz Pedrozo Miguel MG, de Souza Cordeiro C, Silva CF, Pinheiro ACM, Schwan RF. Inoculation of starter cultures in a semi-dry coffee (*Coffea arabica*) fermentation process. Food Microbiology. 2014;**44**:87-95. DOI: 10.1016/j.fm.2014.05.013

[78] Ribeiro LS, Ribeiro DE, Evangelista SR, da Cruz Pedrozo Miguel MG, Pinheiro ACM, Borém FM, et al. Controlled fermentation of semi-dry coffee (*Coffea arabica*) using starter cultures: A sensory perspective. LWT-Food Science and Technology. 2017;**82**: 32-38. DOI: 10.1016/j.lwt.2017.04.008e

[79] Maman M, Sangchote S, Piasai O, Leesutthiphonchai W, Sukorini H, Khewkhom N. Storage fungi and ochratoxin A associated with arabica coffee bean in postharvest processes in Northern Thailand. Food Control. 2021;**130**:1-10. DOI: 10.1016/j.foodcont.2021.108351

[80] Echeverria MC, Nuti M. Valorisation of the residues of coffee agro-industry: Perspectives and limitations. The Open Waste Management Journal. 2017;**10**(1)

[81] Echeverria MC, Pellegrino E, Nuti M. The solid wastes of coffee production and of olive oil extraction: Management perspectives in rural areas. In: Solid Waste Management in Rural Areas. London: IntechOpen; 2017

[82] Aguiar LL, Andrade-Vieira LF, de Oliveira David JA. Evaluation of the toxic potential of coffee wastewater on seeds, roots and meristematic cells of Lactuca sativa L. Ecotoxicology and Environmental Safety. 2016;**133**:366-372

[83] Echeverria MC, Agnolucci M, Cristani C, Battini F, Palla M, Cardelli R, et al. Microbially-enhanced composting of olive mill solid waste (wet husk): Bacterial and fungal community dynamics at industrial pilot and farm level. Bioresource Technology. 2013;**134**:10-16

[84] Dzung NA, Dzung TT, Khanh VTP. Evaluation of coffee husk compost for improving soil fertility and sustainable coffee production in rural central highland of Vietnam. Resources and Environment. 2013;**3**(4):77-82

Chapter 2

The Significance of Blockchain Governance in Agricultural Supply Networks

Michael Paul Kramer, Linda Bitsch and Jon H. Hanf

Abstract

Firms in the agri-food sector have started implementing blockchain technology to both provide transparency over the supply chain transactions and to make trust attributes visible to consumers. Besides the well-known public blockchains such as Bitcoin and Ethereum, private- and consortium-type blockchain platforms exist. The latter ones are being operated in the agri-food ecosystem contributing to the vertically cooperated supply networks that are coordinated by a focal firm. Stakeholders' attitude and behavioral intentions toward the use of the blockchain technology impact their use behavior. The results show that permissioned blockchain governance mechanisms with consensus and incentives to motivate stakeholders are lacking in private and consortium blockchains. This study closes a research gap as understanding how the stakeholder management approach can compensate for the lack of consensus mechanisms can provide managerial guidance toward the development of an effective stakeholder management strategy, which eventually can be provided for a competitive advantage. As there is little research on the role of blockchain as a novel governance mechanism, this research will contribute to the scholarly discussion toward a common understanding.

Keywords: vertical coordination, blockchain, governance, stakeholder management, food industry

1. Introduction

Blockchain technology is a meta-technology that has the potential to change business models and supply chain networks (SCN) in the agri-food industry. Based on a distributed computing architecture, the blockchain protocol in its current form enables the provision of provenance information as well as tracking and tracing to support supply chain management. Recently its implementation in vertically cooperated food supply chains (FSC) has started. FSCs are typically managed centrally with a focal firm being responsible for the coordination of the network [1]. Sensitized by food poisoning cases, consumers nowadays require provenance information and transparency about the production of the food item [2]. Early adopters in the coffee industry are therefore building on blockchain technology (BCT) solutions to provide better visibility about the journey of the coffee product in their supply chain [3]. It has been demonstrated that the application of BCT to provide tracking, tracing and provenance information will result in increased consumer trust [4]. As a result, traceability of food products is becoming a competitive differentiator in the agri-food industry [5].

Since the first Bitcoin block has been minted in 2009 distributed ledger technologies (DLT) such as blockchain have gained rapid acceptance. Oftentimes DLT and BCT are used interchangeably. For purposes of this article, we will continue using the publicly used term "blockchain" respectively "BCT", although blockchain is a category of DLT. We will further use this term synonymously to describe a decentralized, public, and permissionless network. BCT has the potential to develop into a general-purpose technology with the outlook to fundamentally change the economy and society alike. Its decentralized and distributed digital ledger enables secure and trustful peer-to-peer transactions. It is the underlying technology for cryptocurrencies such as Bitcoin and also the basis for the tokenized economy, publicly referred to Web 3. Blockchain relies on the participation of many stakeholders in the decision-making process, in contrast to today's governance in hierarchical organized firms.

In general terms governance refers to the rules and processes of a control system that is used to manage and supervise how stakeholders interact within an organization, a firm, a state, or within an information technology (IT) based network. As such, governance can be seen as a form of regulation that supports the achievement of objectives [6]. The rules of governance coordinate decision-making processes between stakeholders. Governance systems provide for risk mitigation and are also implemented in digital ledger technologies such as BCT [7]. Blockchain as a software protocol enables a new governance infrastructure and its decentralized governance mechanisms involve multiple stakeholders rather than a single authority. Its governance is a self-regulating system that is based on a digital network. Until now, a generally accepted definition of blockchain governance has not yet been agreed upon. As in organizations and firms, decision-making and economic incentives are key attributes of BCT governance [7].

The purpose of decentralized peer-to-peer (P2P) networks is to eliminate the central decision-making authority. Rather than exercising authority by assignment, in decentralized networks authority is exercised by the engagement and experience of the stakeholders. Decisions are being coordinated through consensus mechanisms by the participating entities, by a single entity, or a set of a few entities that have been assigned to perform governance tasks. Consensus in broad terms is an agreement that is being made between various parties. The permissionless consensus mechanism in the cryptocurrency Bitcoin provides for verification of transactions stored in a block. As such, consensus is a confirmation on the status of the cryptocurrency network which leads to a subsequent update of all networked ledgers. Consensus mechanisms are therefore the foundation of the digital trust mechanism in BCT.

The type of governance exercised, and the associated consensus mechanism depend on the type of the blockchain technology platform type (BCTPT) with their varying coordination mechanisms [8]. Governance based on consensus mechanisms such as Proof-of-Work (POW) used with Bitcoin or with the current Ethereum platform[1] to validate transactions can be exercised on-chain (algorithmic, mathematical) or off-chain (human interaction, network policies). We will focus in this article on off-chain rather than on on-chain governance as the behavioral intentions, i.e., the perceived likelihood that supply chain stakeholders will exhibit a certain behavior, is key to our research.

---

[1] As the result of a governance decision Ethereum is currently in the process of swapping the Proof-of-Work consensus mechanism to the energy efficient Proof-of-Stake mechanism.

The research is based on an exploratory case study of the premium coffee producer Solino Coffee based in Germany and Ethiopia. The case study has been selected by purpose due to its uniqueness and because it provides for insights about the influence of instrumental stakeholder management on technology adoption behavior.[2]

The aim of this research is twofold: first, we analyze how the stakeholder management approach impacts use behavior of stakeholders towards adoption of blockchain technology in the absence of permissioned consensus mechanisms in private and consortium BCTPT in vertically coordinated Food Supply Chain (FSC) networks. We want to gain an understanding, how the stakeholder management approach compensates for the lack of involvement in the consensus process. Second, we provide an analysis of the factors influencing blockchain technology adoption behavior of stakeholders in vertically coordinated FSC networks. The article is organized as follows: part 2 explains the research methodology we employed. In part 3 we elaborate on vertical coordination in the agri-food supply chain based on network theory. The theoretical foundation of the research is further built on instrumental stakeholder theory, as well as on technology adoption theory. Part 4 provides for a general blockchain technology overview and discusses the different blockchain technology platform types. Part 5 introduces governance in blockchain and refers to the different governance types that are being exercised. In the subsequent part 6 we introduce the model that determines the technology adoption behavior of stakeholders in the coffee supply chain. Part 7 describes the coffee supply chain case study our research is based on. With part 8 we provide a summary of the results and a discussion on our empirical findings. Eventually, we conclude the article with part 9 and provide guidance in terms of possible directions of future research. The objective is to investigate on the following research questions:

RQ 1: Which governance types apply to the different blockchain technology platforms?

RQ2: How is stakeholder engagement affected by governance mechanisms?

RQ3: How does the stakeholder management approach impact use behavior of stakeholders?

2. Methodology

The research methodology we followed is based on three pillars: an extensive literature review concerning BCT in vertically coordinated agri-food supply chains including network and technology adoption theory, a quantitative survey, and expert interviews. In addition, we have been, and we still are involved, in ongoing discussions with operators of BCT solutions. These operators have first practical experiences with the operationalization of BCT for the purpose of provision of tracking and tracing as well as of provenance information in agri-food SCNs. Based on the theories we have constructed a blockchain technology adoption model. Our proposed model combines principles of technology adoption, economics, and social psychology to investigate the behavioral intention of individual stakeholders in the coffee FSC towards BCT adoption.

Empirical data has been derived from an explorative case study of the premium coffee producer Solino Coffee based in Germany and Ethiopia which provides empirical insights into the applicability of the proposed model of technology

---

[2] The authors of the current paper contributed to the case study-based article "Nothing else matters: Blockchain technology adoption behavior of stakeholders in rural areas" which has been submitted in June 2021 [9].

adoption behavior. The production facility of Solino Coffee is located in the Ethiopian capital of Addis Abeba. As a few early adopters in the coffee industry have just implemented and operationalized BCT to enhance their supply chain, this is the earliest possible point in time to conduct research with the objective of obtaining meaningful results. Quantitative data has been collected through online interviews and qualitative data through expert interviews. Our online questionnaire methodologically follows the Reasoned Action Approach [10]. The questionnaire is based on five factors determining usage behavior and behavioral intention.

This case study has been selected by purpose due to its uniqueness and because it provides insights into the influencing factors of instrumental stakeholder management on technology adoption behavior.

3. Literature review

3.1 Attributes of agri-food supply chain networks

Food supply networks in the agri-food business are typically managed centrally with a focal firm being responsible for the decisions relating to the coordination of the network [11]. The networks have been classified as strategic networks where the focal firm is responsible for all attributes of the food item in the network [11]. Attributes of strategic networks are the hierarchical coordination through a focal firm, the intensity of relations, and the coordination mechanisms. Strategic networks are mainly organized in a pyramidal-hierarchical structure, in which a focal firm acts as the centralized decision-making authority which coordinates the network. The focal firm also sets the strategy and aligns the actions in the network [12]. Strategic networks are characterized as long-term relationships of power and trust through which organizations exchange influence and resources between at least two or more actors in the network. Furthermore, strategic networks can be seen as a construct of long-lasting inter-organizational links which have a strategic significance for the participating firms [13]. Gulati showed that coordination and cooperation are two important means for the management of vertical relationships [14].

3.2 Coordination mechanisms

Coordination can be understood as the alignment of actions to mutually achieve goals between intentionally chosen partners. Coordination problems arise if actors are not aware that their actions are interdependent or that there is uncertainty about others' rationality so that one does not know how others will act. Thus, problems of coordination are the result of the lack of shared and accurate knowledge about the decision rules that others are likely to use, and how one's actions are interdependent on those of the others [15]. There are formal (including programming, hierarchy, and feedback) and informal mechanisms (leadership, norms, culture, shared values and experience, trustworthiness, and a shared strategy) to overcome coordination problems [16]. Cooperation refers to the alignment of interests between the partners in which the intended scope of the relationship is laid out. Thus, problems of cooperation accrue from conflicts of interests because self-interested individuals optimize their benefits before they strive for collectively beneficial outcomes [13]. Formal mechanisms such as contracting, common ownership of assets, monitoring, sanctions, prospects of future interactions and informal mechanisms such as identification and embeddedness can be used to overcome the problem of motivation.

Coordination mechanisms in food supply chain management can be broadly divided into six groups: power, contractual relationships, information sharing, joint decision-making, collective learning, and building routines [17, 18]. In addition, Pietrwicz examined consensus building as well as coding and executing smart contracts as coordination mechanisms for online transactions [19]. Cooperation and coordination therefore need to be seen as two indispensable parts of the supply chain collaboration. For an efficient management of vertically cooperated FSC, it is necessary to manage both mechanisms simultaneously [1] where the key objective of the participating supply chain stakeholders is to provide the end customer with the products and services that are being demanded.

3.3 Blockchain as a trust technology

Vertical cooperation in the food industry is driven by trust attributes such as food quality, provenance, and safety [11]. In a pyramidal–hierarchic organization, decisions are made by the focal firm, which is responsible for the strategic direction of the SCN. According to Ketchen and Hult, intermediaries and agencies in supply networks increase the potential for abusing power and intentionally take advantage of the SCN, which is the result of a single decision authority [20]. According to Belaya and Hanf power can be used as an effective coordination mechanism in the FSC operating in centralized ecosystems [21]. However, power could also be applied to the advantage of the network to solve issues and problems in supply networks [22]. The level of decision-making power applied to the supply network is critical for its efficiency, with a higher degree of control resulting in an increase in supply network value. It has also been proven to impact the management of highly interconnected networks, where the supply network performance suffers less with higher control applied [23]. A trust attribute of BCT is that blockchain is consensus safe as transactions can only be executed when the majority of participants verify them. Consensus is an agreement that is being made between various parties when participating entities reliably and efficiently verify transaction attributes [24]. The decentralization nature of a blockchain system impacts the level of control as well as the decision-making [24]. BCT is performing decision-making through consensus about the contents and validity of transactions as an aspect of coordination. In a decentralized network, decisions are made by the joint consensus of the participating entities. In public BCTPT decision-making is typically performed through mass consensus. In private BCTPT a single ruling entity performs the decisions alone; in consortium platform types authorized participants perform the decisions. As a result, based on the BCT platform, different consensus algorithms apply [8]. The founder of Ethereum blockchain, a public BCTPT provided a detailed explanation of the consensus algorithm: *"The purpose of a consensus algorithm, in general, is to allow for the secure updating of a state according to some specific state transition rules, where the right to perform the state transitions is distributed among some economic set. An economic set is a set of users which can be given the right to collectively perform transitions via some algorithm, and the important property that the economic set used for consensus needs to have is that it must be securely decentralized - meaning that no single actor, or colluding set of actors, can take up the majority of the set, even if the actor has a fairly large amount of capital and financial incentive."* [25].

BCT is a trust technology through the application of its attributes of transparency, integrity of data, and immutability [26]. It enables the sharing of product trust attributes with consumers, making supply chain activities more transparent [3, 27]. BCT with its trust attributes consensus, immutability of data, cryptographic security, and transparency could create a new trust platform for business transactions as the application of disruptive technologies such as BCT to the agri-food supply chain management can increase trust by generating closer relationships between the firms [28].

3.4 Instrumental stakeholder theory

As part of his instrumental stakeholder theory Freeman defines a stakeholder as "any group or individual who can affect or is affected by the achievement of the organization's objectives" [29]. Freeman views his theory to be used in the realm of management's strategic decision making. The traditional instrumental stakeholder theory focusses specifically on independent, dyadic relationships whereas a newer strand argues that organizations are represented by a complex network of horizontal and vertical relationships [30, 31]. Following the instrumental stakeholder theory successful cooperation between management and stakeholders provides for a competitive advantage. Understanding the factors influencing behavioral intentions of stakeholders while using BCT can provide guidance as to what extent management can support and motivate stakeholders in using the technology, which eventually can provide for a more efficient use of the technology. This in turn could result in a distinct competitive advantage for the firm. When management introduces new technologies, they can be faced on one hand with the challenges that accompany the pure technical implementation but on the other hand more importantly they can be confronted with stakeholder resistance, unwanted attitudes towards usage, and potential anxiety of the users. The latter is the reason why we chose to analyze the behavioral intentions of stakeholders towards adopting and using BCT and putting this into perspective to the chosen stakeholder management approach, especially against the background of the novelty level of BCT. Employees in an organization can use their power and resist to changes through forms of behavior that do not support the objectives of the organization. It is therefore imperative that management must be aware of the stakeholders' attitudes and behavioral intentions towards the usage of new technology. Lazzarini et al. show that the normative path of stakeholder theory can lead to a strong commitment of the organization in adhering to the strategies that have been set by management [32]. The normative view of stakeholder theory focusses on the state that should be achieved. Management and stakeholders therefore need to take each other's objectives, motivations, intentional behaviors, and concerns into account to jointly strive for the envisioned economic rent of the firm. Consequently, management has to ensure that affected stakeholders accept and adopt the novel technology in order to achieve the expected economic rent. Stakeholder theory has been argued to be descriptive, which is the collaboration amongst stakeholders, instrumental, which assesses stakeholder management conduct and supply chain performance, and normative, which describes the attitudes of the firm towards its stakeholders. All three attributes support each other and are based on a normative foundation focusing on the value of economic fairness and corporate social responsibility (CSR) or on factors determining what an economy should represent [33]. To be economically successful and outperform their peers, firms should also enter into contractual agreements with their stakeholders following the instrumental stakeholder theory [34]. This coincides with the strategic value chain approach which views the value chain as a single solution improving the competitive position by putting the customer and their expectations first, to improve the overall chain performance [35]. Consequently, the development of close ties between the firm and its stakeholders has the potential to result in sustaining competitive advantage [36]. As the actions of stakeholders in the supply chain affect the value of an asset, BCT must be accepted, adopted and used by users to gain productivity advantages [37]. Stakeholders should have control over the asset e.g., over the BCT, to maximize their utility and satisfaction. Stakeholder theory asks managers to understand the needs, motivations, and interests of stakeholders and also factor in their experience and skills to increase the supply chain efficiency [38]. Stakeholder theory can be applied to IT projects and will be effective in that industry [39]. As blockchain is a software protocol that is being implemented with the IT infrastructure it can be viewed as an information technology asset [40].

3.5 Technology adoption

Technology adoption can be viewed from an organizational or an individual stakeholder level [41]. For the purpose of this article, we analyze user adoption from a blockchain user perspective and focus on these stakeholders that have been tasked by their principal to add data to the blockchain ledger. We utilize the Unified Theory of Acceptance and Use of Technology (UTAUT) which is amongst the three most common models to analyze technology adoption and use behavior of information technology [42]. UTAUT has been used in numerous studies to analyze and predict the acceptance and adoption of technologies predominately from the user perspective. It is based on four factors determining usage behavior and behavioral intention: performance expectancy (PE), which is the support of the technology for achieving the individual's objectives, effort expectancy (EE), which relates to the level of how easy an application is to be used, social influence (SI), which is the perceived influence of others to use the technology and facilitating conditions (FC), the support of the organization towards the individual using the technology. Complementing UTAUT, the Theory of Planned Behavior (TPB) predicts behavioral intent of individuals and the consequences of their behavior [43]. TPB has been built on three independent factors of intention, which are attitude (AT) towards the behavior and answering the question of whether the use of the technology will make a positive difference, subjective norm (SN), which investigates the perceived peer pressure to use a technology, and perceived behavioral control (PBC), which answers the question if the user has the appropriate tools to be successful. Those three independent factors of intention make up the believes of an individual which in turn drives their social behavior [43].

3.6 Trust as the Foundation of Economic Activity

While human trust is being exercised on the social and economic level, digital trust is being exercised on the crypto-technology level. The combination of both trust levels enables the development of novel business models. The increased demand in FSC transparency initiated a redesign of the food chain which is driven by trust attributes such as product quality and food safety [11]. Consumers are increasingly demanding a high level of product quality and safety and expect transparency about their food products, including information about provenance, suppliers, production, and transport conditions [2]. Trust has become a significant element of product quality and safety for which the focal firm is standing in with its brand to constantly ensure high standards. Consequently, agri-food firms need to provide food product related information with the objective to increase trust which could increase customer loyalty and which in turn offers the opportunity to convert one-time buyers to repeat buyers. Trust attributes in the FSC can be split into three categories: the metaphysical, chain transparency, and risk-related category [11] and can be used as a differentiator to enforce price premiums [44]. Examples for metaphysical, non-sensory credence attributes for the coffee industry are including but not limited to coffee completely produced at the place of origin [45] and coffee that has been hand-picked or hand-picked exclusively by women [46]. In our research we combine metaphysical and chain transparency trust attributes. Trust is also a central driver for achieving collaboration in vertical cooperation [46]. It is instrumental in managing the risk of cooperation problems in FSC [1]. In the FSC trust has the potential to reduce transaction costs while fostering cooperation [47]. Previous research has shown that trust has a positive effect on agricultural stakeholder's technology adoption efficiency [48].

4. Blockchain platform types

Blockchain is a decentralized and distributed digital ledger, enabling secure and trustful peer-to-peer transactions. It is the underlying technology for cryptocurrencies such as Bitcoin and also the basis for the tokenized economy, publicly referred to Web 3 [49]. The most prominent blockchains such as Bitcoin or Ethereum are public and permissionless.

BCT as it is being implemented today in agri-food supply networks can be viewed as an institutional technology as it is "a new institutional technology of governance that competes with other economic institutions of capitalism, namely firms, markets, networks, and even governments" [50]. BCT is revolutionizing governance and adhering to Williamson's New Institutional Economics theory, BCT is an institutional technology [51, 52]. This applies to blockchain in vertically cooperated supply networks where provenance and track and trace solutions dominate. Blockchain in permissionless public networks potentially develop further into a general-purpose technology (GPT) with the outlook to fundamentally change the economy and society alike creating a new type of economy [53–55]. As the change in governance is key to our research, we will follow the institutional view of Davidson et al. and view BCT as an institutional technology utilizing aspects of the transaction cost theory.

4.1 Blockchain technology

Blockchain is based on the distributed ledger technology (DLT), a constantly synchronized ledger distributed across locations and entities. As it comprises of various existing technologies that are, intelligently combined, creating a new technology It can be viewed as a meta-technology [56]. In addition to DLT, certain blockchain solutions have been designed to set up rules for transactions enabling the development of decentralized applications, smart contracts, and digital autonomous organizations (DAO). Smart contracts are software programs that are based on BCT with rules for automatically executed transactions based on a set of predefined conditions that have to be met [57] whereas DAOs are a combination of several smart contracts executing on pre-determined business processes. Smart contracts in BCT can be seen as coordination mechanisms applying an institutional perspective over coordination [58]. A fungible token, the digital, alphanumerical representation of a physical asset such as a Bitcoin, is the simplest form of a smart contract. One of the key characteristics of blockchain is the decentralization of the network architecture enabling peer-to-peer transactions, which eliminates the need for a coordinating trusted entity. Trust is induced through the consensus algorithm, the ubiquitous visibility of transactions, the immutability of the data, and the anonymity of trading entities. The trust of the central authority is replaced by the consensus algorithm as transactions can only be executed when participants in the network approve them. The self-organizing peer-to-peer data-sharing technology operates without a central authority or intermediaries.

Although BCT is not managed by a central authority BCTPT exist that provide for a centralization of control [39]. Three different platforms exist today: the public, private and consortium BCTPT. They are predominantly differentiating through access rights and their rights to read from and write into the ledger. What all BCT platforms have in common is the distributed ledger technology, peer-to-peer transaction capability, as well as a generic consensus mechanism. However, different governance types apply to the BCTPT.

4.2 Public Blockchain platform

Decision-making in the way of verifying transactions in blockchain systems is performed through consensus. The consensus protocol ensures that participating entities agree on adding new transaction data to the blockchain replacing the central authority. In the public BCTPT consensus is typically achieved through the majority of the participating entities utilizing for example the Proof of Work (POW) or Proof of Stake (POS) consensus algorithm. The public network is open for participation to everyone and everyone has access and visibility to the transaction data in the ledger, can verify transaction blocks, and participate in the consensus process. Nodes can be added by anybody without the permission of a central authorizing entity as the only requirement is an internet connection and a computer platform. This type of BCTPT is called permissionless as no permission from an authority is needed to participate in the network. Transaction data, once verified, is secure and immutable.

Governance of public blockchains combines decision-making processes with incentives to secure the long-term operation of the network. In public networks, which operate the POW algorithm that is being used with Bitcoin, miners receive incentives for finding the nonce (number only used once). The prospect of the rewards drives miner's behavior to keep minting transaction blocks which are cryptographically hashed to ensure the immutability of the transaction data. The nonce is used to calculate a block hash that meets specific requirements. The first miner who finds the nonce resulting in a valid hash for the block, receives cryptocurrency as a reward. The validity is being confirmed by the participating entities operating blockchain nodes. The combination of technical as well as economic effects account for the proper functioning of the network. POW and POS are well known consensus algorithms that are being used in public BCTPT.

4.3 Private Blockchain platform

In a private BCTPT a single central authority is responsible for the decision-making process and is therefore performing the governance task of consensus building. A single ruling authority is coordinating the permissioned access and verification of transactions. Private BCTPT are mainly used in enterprise environments. There, the central authority approves the access of entities that are permitted to participate in the network. As the decisions are being made by a central authority network consensus remains in one hand. As a consequence, transactions are being verified much faster and transaction throughput can be much higher compared to public BCTPT.

4.4 Consortium Blockchain platform

The consortium BCTPT is also a permissioned technology such as the private type, as only authorized participants will be granted access to the network. In contrast to the private platform the network is being controlled by a group of entities having equal voting rights and jointly operating and maintaining network and system technology. In contrast to private BCTPT, delegated participants perform the decision-making process, authorized to perform consensus building. The system is decentralized, and its aim is rather collaboration than competition between the participating firms. Cost savings, accelerated learning, and sharing risks are the top benefits organizations expect from a certain consortium according to a recent research conducted by Deloitte [59]. PBFT (Practical Byzantine Fault Tolerance) and PoA (Proof of Authority) are examples of consensus algorithms executed in private and consortium BCTPT.

5. Decentralized network governance

Corporate governance is the factual and legal regulatory framework of firms to exercise good corporate management practice. It combines control and monitoring activities while striving for adhering to the economic and social objectives as well as the interests of their stakeholders [60]. In general terms governance refers to the rules and processes of a control system that is used to manage and supervise how stakeholders interact within an organization, a firm, a state, or within an IT-based network. As such, governance can be seen as a form of regulation that supports the achievement of objectives [6]. The rules of governance coordinate decision-making processes between stakeholders. Governance systems provide for risk mitigation and are also implemented in digital ledger technologies such as blockchain [13].

In contrast to the neo-classical approach, transaction cost economics (TCE) assumes that human beings are not capable of making perfectly rational and logical decisions [61], although this has been implicitly presumed in previous studies [62]. Human's decisions are limited by their cognitive abilities including but not limited to processing large amounts of data, their emotions, and the limited amount of time they have for making decisions without exploring all available alternatives or obtaining all relevant information which results in decision making based on incomplete information. Hence, humans are not able to make perfectly rational and logical decisions according to Herbert A. Simon's theory of bounded rationality [63]. As per TCE humans also act opportunistically, seeking to enforce their strategic objectives. Replacing human's limited decision-making capabilities by information technology solutions has the potential to impact bounded rationality and opportunism.

Entering into contractual agreements with other firms is associated with costs which are defined as transaction costs. Transaction costs, as result of the coordinating activities, can be ex-ante, which are costs associated with information gathering and searching for the right partner and cost of negotiation and entering into a contractual agreement or ex-post, which are costs associated with overseeing the transactions according to the agreement and applying the necessary measures if the transactions deviate from the contractually agreed framework. As transaction costs deriving from bounded rationality and opportunism of contracting parties diminishes the integrity of contracts, transactions should be organized so "as to economize on bounded rationality while simultaneously safeguarding them against the hazards of opportunism." [50]. To safeguard against the challenges of bounded rationality, opportunism and information asymmetry governance mechanisms are employed to maintain an orderly transaction process, reduce conflicts, mitigate problems to ensure profitable transactions.

Governance mechanisms can be described as reactions to incomplete knowledge, uncertainty, dependence, and opportunism between firms [51, 64]. From a

TCE perspective trust and incentive are governance mechanisms [65]. Trust should safeguard against the risk of opportunism to ensure efficient coordination of transactions [65], whereas incentives are governance and control mechanisms used to coordinate the interests between principals and agents while at the same time reducing agency-related challenges [65, 66]. Firms determine on the governance type to benefit from the gains of cooperation and coordination [64, 67].

Governance processes can also be executed in decentralized networks without the need of a trusted central authority [48]. This is the key concept of BCT. As a trusted entity is missing, the network has to ensure that decisions are being made in such a way that while performing transactions and transferring an asset it has to be prevented that a digital copy of the asset is being transferred multiple times which refers to solving the so-called double spending problem [68]. The economic problem of double-spending has been solved in decentralized networks with the implementation of multiple nodes carrying identical ledgers where consensus mechanisms such as POW or POS apply. Blockchain as a software protocol enables a new governance infrastructure and its decentralized governance mechanisms involve multiple stakeholders rather than a single authority. Blockchain governance is a self-regulating system that is based on a digital IT network. Research on blockchain governance is still scarce [69]. Until today, a generally accepted definition of blockchain governance has not yet been agreed upon.

Blockchain governance differs significantly from traditional corporate governance and can be seen as a new form of organizing collaboration between firms [70]. Lumineau et al. investigate blockchain governance from a meta-perspective and conclude that governance in blockchain can be viewed distinct from the traditional mechanisms of corporate both contractual and relational governance [68]. Contractual governance defines the control mechanisms and rules for enforcing legal contracts, relational governance addresses behaviors, the joint value system and prospects of future cooperation. The application of governance mechanisms for the verification of transactions is key to the operation to any blockchain network. Along the lines of Douma's definition that "Corporate Governance is the system by which business corporations are directed and controlled" [71], control in different BCTPT exercised through governance mechanisms is an aspect of the governance type. Where the intensity of control in vertical coordinated FSCs moves along a continuum ranging from spot market to vertical integration, the blockchain continuum ranges from no control in public BCTPT to single control in centralized systems. The control intensity in productive partnerships is beneficial to farmers and processing companies alike [72]. Farmers in FSCs benefit from a level of control that its being exercised through contractual agreements that amongst other benefits secure their income, enables production planning, and education [73]. We therefore hypothesize that control in blockchain governance can be described as blockchain governance continuum framework. The continuum suggests that the intensity of control exerted is developing from no control at public BCTPT to truly centralized implementations with sole control by a single authority. The continuum has been summarized in **Figure 1**.

Despite the decentralized character of public BCTPT certain consensus protocols such as POW in public networks support centralization efforts, as miners could agree and collaborate to achieve a 51% share of all mining activities. With his behavior, false transactions could be verified despite the fact that decentralized networks should be safeguarded from manipulation.

5.1 Trust through consensus

In the economy trust is needed to utilize assets such as gold, shares, or Fiat money as a store of value. Pass et al. define the store of value as any asset that can

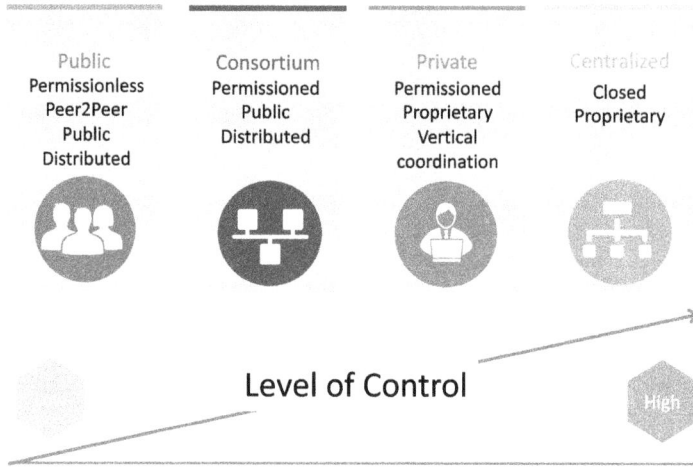

Figure 1.
Framework blockchain governance continuum.

be converted back money with a positive difference between purchase and sell price [73]. Using money for transactions is based on a consensus in its role as store of value. Trust in an asset such as Fiat money, which is the most common store of value, is generated by the issuing central authority, which is typically an assigned entity in the country that manages money supply.

A major security and trust factor in BCT is the process which adds transaction blocks to the blockchain. An algorithm ensures that the network consents on the chronologically sorted and constantly growing set of blocks. As part of this algorithm the specific consensus algorithm ensures that verification nodes agree on the validity of the block to be added. In a blockchain system transactions are being verified according to the governance type that has been chosen to operate the decentralized network. BCT establishes trust by using consensus mechanisms for decision-making. The consensus-based verification process establishes the trust in the transactions and the incentives drive the behaviors of the stakeholders.

The consensus mechanism in BCT is an integral part of its architecture and has been embedded as a separate layer in the layered blockchain architecture. A simplified overview of the BCT architecture layers is presented in **Figure 2**. Consensus acts as a confirmation on the status of the network which leads to a subsequent update of all networked ledgers. The consensus mechanism is therefore the foundation of the digital trust mechanism in BCT.

5.2 Consensus mechanisms in blockchain platforms

In our article we will analyze the governance types used by different BCTPTs and how the governance mechanisms affect coordination of the network. Decision-making and economic incentives are governance categories both, in organizations and firms as well as in BCT [13].

Public and permissionless blockchains operate under the governance type of public consensus as transactions can be verified by any participating node. The access to the network is permissionless which allows everyone with a computer and an internet connection to join the public network. The consensus algorithms such as POW or POS also operate permissionless. As such, the governance categories as decision-making and incentives through consensus are driving the behavior of stakeholders.

Application Layer (dec. apps, BaaS, APIs, SDKs)

Consensus Layer (different algorithms)

Ledger Layer (blocks, Merkle trees)

Peer-to-Peer Layer (transaction, token, smart ctrct.)

Network Layer (TCP/IP, centr., decentr., distributed)

Hardware Layer (clients, servers, nodes, IoT devices)

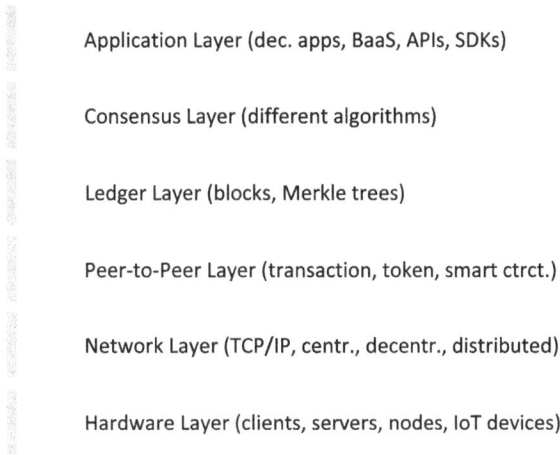

Figure 2.
Layered blockchain architecture.

In private BCTPTs the consensus is coordinated by a single central authority. The consensus algorithms deployed in those platforms therefore differ from those of public BCTPTs. Examples for consensus engines in private networks are Tendermint or Practical Byzantine Fault Tolerance (PBFT). They predominately ensure that transaction blocks are chronologically stored on each participating node. The governance type in private BCTPTs is reverted back to single management of transactions.

The governance type used with consortium BCTPTs differs slightly from that of private ones. The key difference is that the consensus is coordinated by a few, assigned entities who are authorized to perform governance related tasks. Same consensus engines are used as in private BCTPTs. As a result, different governance mechanisms apply for the three BCTPTs.

While public blockchains permit any entity to become stakeholder and participate in the network governance, permissioned blockchains such as private and consortium BCTPT are lacking the governance mechanism attributes of public consensus and incentive as either a single authority or a few pre-determined stakeholders are managing the consensus algorithms with permission. As a result, in private and consortium BCTPT the majority of the stakeholders are excluded from contributing to achieving consensus on transactions and on the state of the blockchain network. As a consequence, alternative governance mechanisms need to be applied in private and consortium BCTPT as consensus and incentive mechanisms are no longer available as governance attributes.

5.3 Off-chain vs. on-chain governance

As part of corporate governance, stakeholder management is being tasked to balance the different interests and motivations of stakeholders. The theory is based on the concept of a hierarchical organized enterprise. Although a blockchain network has different stakeholders the objective of blockchain governance is the same: to balance the interests of stakeholders. Katina et al. conclude that blockchain governance is based on three pillars which are direction, oversight and accountability where consistent decision-making is an attribute of direction, control of the system an attribute of oversight, and performance regarding the monitoring of resources and the system that of accountability [74]. The objectives of stakeholder management are similar in centrally coordinated and decentralized networks. However, different stakeholders are acting in both network types, further also depending on the chosen BCTPT. In selecting a blockchain governance type it is vital to analyze the various stakeholders involved in the network, the incentives that can be achieved, and the coordination of the stakeholders to conclude on valid transactions.

Blockchain stakeholders include but are not limited to contributors, steering board members, software developers, miners, platform operators, and users. Contributors are engaged in the funding of blockchain projects taking also a consultative role. Their incentive is the initial stake that they hold in the venture. Members of the steering board are consulting and collectively deciding on the future direction of the BCT. Their incentive is the potential gain of their stake in the network. Software developers are hardcoding rules for the BCT protocol. Other than a contractual relationship and potentially an early stake in the technology there is no incentive mechanism in place for them. While the consensus mechanism represents the DNA, miners represent the heart of the network. They ensure a constant flow of transactions that need to be verified. As incentive they receive either transaction fees or block rewards for hashing a transaction block. Node operators participate in the verification of transactions and benefit from the collectively gathered and distributed transaction fees. Platform operators are responsible for the provision of the IT infrastructure blockchain applications are running on and users are stakeholders that create transaction data and use the blockchain to achieve economic gains. Users pay for the transactions they initiate and for the use of the network service. Users are also adding data as agents of the principal that has tasked them to do so.

Decisions relating to transactions and to the operation of the network can take place off-chain on a social level where decisions impact the architecture, software code, processes, and consensus mechanisms in the blockchain as well as on-chain on a technical level where pre-coded algorithms that are implemented in the blockchain protocol perform tasks according to the predetermined rules. Both, the scientific and public literature describe on- and off-chain governance differently. As no common definition exists, we propose the following wording:

> *"Off-chain governance refers to the rules and decision-making processes, the communication between the involved stakeholders and the future development of the blockchain code. On-chain governance refers to the software-coded algorithmic enforcement of rules in the decentralized network concerning changes to the blockchain protocol, block verification, decision-making, and reward mechanisms."*

For the purpose of this research, we focus on off-chain governance mechanisms as the objective of the research is to analyze how, in the absence of mass consensus mechanisms, the consensus mechanism is being compensated for in private and consortium BCTPT through a stakeholder management approach.

6. Technology adoption research model

Coffee supply networks are strategic networks and both private and consortium BCTPT are supporting the supply chain management. Based on UTAUT and TPB we have constructed a model to analyze the intent of users in the coffee supply network towards BCT adoption to predict the individual behavior of the stakeholders. Our proposed model combines principles of technology adoption, economics, and social psychology to investigate the behavioral intention of individual stakeholders to adopt BCT.

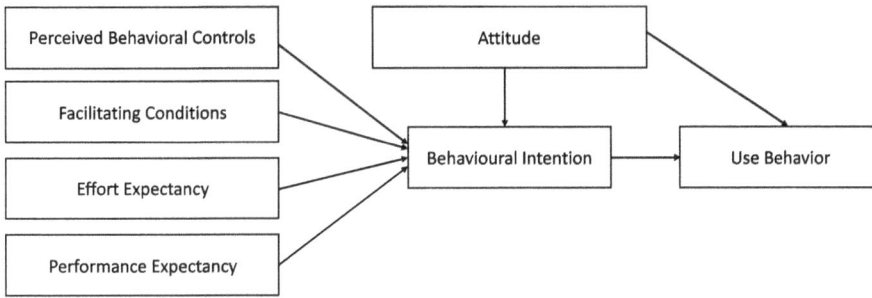

Figure 3.
Proposed theoretical blockchain technology adoption model (source: Authors, based on Ajzen, 1991 [41]).

UTAUT serves as a basis of the proposed model from which we selected the three appropriate parameters PE, EE, as well as FC, which have been supplemented by two additional parameters from PBT, namely AT and PBT. Not all parameters from UTAUT and TPB are applicable to the case study, and we argue why we exclude certain aspects. Theory on technology adoption suggests that older people and women are more sensitive towards social influences [41] and only in the early phase of technology usage as well as in mandatory settings there is significant evidence that the opinion of peers influence subjective norms as a determinant of behavioral intent [75]. There is also still a lack of understanding about the user acceptance or rejection of IT [76]. Consequently, we exclude the determinant subjective norm from TPB when building our proposed model. From UTAUT we excluded the parameter SI because according to Venkatesh it only fits a small target group and could potentially dilute the result [41]. The factor attitude is a key determinant of behavioral intentions and directly influences usage behavior [75]. Attitude plays a key role in the adoption of information technology. We propose the following model which should accurately predict FSC stakeholders blockchain technology adoption behavior (**Figure 3**).

7. Case study

Solino Coffee Products (Solino) is partnership under German law [44]. Coffee products are being produced completely in Ethiopia shifting major parts of value creation to the country of origin. Tasks include sourcing, roasting, packaging, labeling, and coordinating the transport to its German distributor. The business challenge was to provide trusted information about the coffee products in the supply chain in their quest to further increase customer loyalty as consumers are increasingly asking producers to make the supply chain processes more transparent to them; at present, this applies especially to the provenance information about the products sold. Solino is one of the first firms in the coffee industry that started their BCT implementation in 2018 while it is progressively adding more function-ality to the supply chain. Every stakeholder of the Solino supply chain who adds data and value to the business process adds their data to the distributed ledger. In the case of coffee, smallholders, collectors, or cooperatives enter data about the date of harvest and where the coffee was harvested. Further information about the transfer of goods, roasting, and shipping are being recorded in the blockchain ledger. BCT provides benefits for both sides: Solino provides consumers with access to provenance information while the management of Solino is transparently monitoring their supply chain activities: from harvesting to roasting and shipping to Hamburg.

Solino has chosen a normative stakeholder management approach for their business operations in Ethiopia. The company and its stakeholders are jointly striving towards creating the majority of added value in the country where the raw material, the coffee cherries, originates. The interests of the stakeholders are dominating the business conduct rather than focusing on the economic rent of the firm. This approach emphasizes morals and ethical conduct while displaying a high degree of corporate social responsibility (CSR).

8. Results and discussion

We analyzed governance types and consensus mechanisms in public, private, and consortium BCTPT answering the first research question. It was demonstrated that public BCTPT permits any entity to participate in the network governance and transactions can be validated by every participating entity. Governance is exercised through public consensus mechanisms. Permissioned blockchains such as private and consortium BCTPT are managing the consensus algorithms with permission. The governance mechanism attributes of public consensus and incentive are not available for users of those BCTPT. As a result, in private and consortium BCTPT the majority of the stakeholders are excluded from contributing to the state of the blockchain network which answers the second research question. The corporate governance model applies to those BCTPTs.

Blockchain is a network developed and maintained by humans. Motivational psychology describes the two stimuli of humans' behavior which can either be extrinsic or intrinsic. Motivational psychology describes the two stimuli of humans' behavior which can either be extrinsic or intrinsic. In the absence of governance mechanisms extrinsic or intrinsic motivation factors need to be in place so that humans get incentivized to contribute to the operation of the network. Extrinsic motivation refers to the achievement of an objective that is driven by a reward despite the fact that the individual does not prefer to perform the action. It is therefore instrumental as the result is separated from the objective. Intrinsic motivation is stimulated from within the individual because the behavior exercised is naturally rewarding and satisfying.

Off-chain governance in private and consortium BCTPT is exercised through the stakeholder management approach. In order to understand how the stakeholder management approach impacts use behavior of stakeholders towards adoption of blockchain technology in the absence of permissioned consensus mechanisms we conducted on online-survey with blockchain users of Solino Coffee. In response to our third research question that asks for the factors that impact the adoption of blockchain technology by stakeholders in the coffee supply chain we developed a blockchain technology adoption model and interviewed stakeholders in the upstream coffee supply chain including coffee roasters, packing specialists, quality managers, as well as logistics managers. We addressed research questions of how these factors impact coffee supply chain performance by unveiling that normative stakeholder management obviously positively influences technology adoption behavior. Our findings further unveil that close ties between management and stakeholders positively influence behavioral intentions and subsequently the usage behavior of stakeholders towards blockchain technology adoption. Our findings suggest that the application of a normative stakeholder management approach coincides with strong positive behavioral intentions and strong positive usage behavior and ccompensates for the lack of consensus mechanisms in private and consortium BCTPTs. Our findings express a consistently high level of the attitude factor amongst stakeholders in the upstream portion of the supply chain towards adopting BCT. Stakeholders view the adoption of BCT as a critical success factor which affect them personally as the use of BCT will make a positive difference in their job and future career development. In addition, stakeholders strongly confirm through the PBC and FC factors that the enterprise is providing the appropriate IT tools to be successful. Stakeholders exercise a high belief in the technology to support them in achieving their individual job objectives. The results of the interviews also highlight the importance of EE which refers to the ease of use of the application that is driving the behavioral intention. AT and PBC are the strongest influencers of behavioral intentions which drive usage behavior. As per our model, attitude is directly impacting behavioral intentions as well as usage behavior which are key determinants of adopting IT technology. AT and PBC factors strongly impact BCT adoption behavior of stakeholders the agri-food supply chain. PE, EE, and FC conditions also impact the adoption but with a less strong characteristic. We also found that PE, EE, FC, PBC, and AT positively influence the usage of BCT in the coffee production process independent of age, gender, job function, and professional experience. We conclude that BCT adoption has a mediating role and is one of the key factors affecting supply chain performance.

We admit that the chosen research methodology has certain disadvantages including but not limited to the single case study, the data obtained through interviews and that the findings can only be applied to this specific case. As BCT is a novel technology, research on adoption of BCT by stakeholders in the supply chain has just provided some preliminary and limited research to date examining this topic. In order to overcome this shortfall, in addition to the case study we have conducted interviews with industry experts with experience in similar implementations with the objective to scale our single case study findings.

9. Conclusion

This article investigated how consensus mechanism is being compensated for in private and consortium BCTPT through a stakeholder management approach impacting stakeholder's use behavior towards the adoption of blockchain technology. The results show that permissioned blockchain governance mechanisms with stakeholder consensus and incentives implemented to motivate network stakeholders are lacking in private and consortium blockchains. The blockchain technology platform types are exercising different governance types with associated consensus algorithms.

We combined secondary and primary research to find evidence that the choice of a normative stakeholder management approach can replace the blockchain governance in private BCTPT, thus positively influencing the behavioral intentions of stakeholders and subsequently their usage behavior towards blockchain technology. It has been shown that the lack of blockchain governance in private BCTPT is compensated for by intrinsic motivational factors. This study closes a research gap as understanding how the stakeholder management approach can compensate for the lack of consensus mechanisms can provide managerial guidance towards the development of an effective stakeholder management strategy, which eventually can provide for a competitive advantage.

Considering that the research is based on a single use case, the individual circumstances need to be taking into account when applying the results to other supply chains. The decision on the most beneficial governance type needs to be carefully analyzed on a case-by-case basis.

We acknowledge that our study has some limitations as it is based on a case study which is subject to individual interpretation of the authors. We mitigated this shortfall by adding qualitative data about a similar case study. The blockchain technology has also just been operationalized in some coffee supply networks and this is the earliest possible time to obtain meaningful results derived from stakeholders of the blockchain supply chain. Our study also leads to future areas of blockchain governance research. Subsequent research needs to expand on the research object and include consortium BCTPT. Novel categories of potential blockchain governance mechanisms such governance tokens that represents decision-making capabilities referring to blockchain protocol implementations need to be included as well. Smart contracts have governance mechanisms hard coded in their application and automatic transactions play a key role in the future blockchain implementations. Special focus needs to be put on the role of fungible and non-fungible tokens. At the end, off-chain and on-chain governance mechanisms are originated by humans and it needs to be investigated how the application of Artificial Intelligence with its Deep Learning capabilities could impact on-chain governance.

Author contributions

Michael Paul Kramer developed the theoretical formalism and wrote the article. Linda Bitsch and Jon H. Hanf contributed to the design and implementation of the research, to the analysis of the results and provided feedback on the writing of the final version of the article. Jon H. Hanf also supervised the project. All authors have read and agreed to the published version of the manuscript.

Funding

This research received no external funding.

Conflicts of interest

The authors declare no conflict of interest.

Author details

Michael Paul Kramer*, Linda Bitsch and Jon H. Hanf
Hochschule Geisenheim University, Geisenheim, Germany

*Address all correspondence to: michael.kramer@hs-gm.de

IntechOpen

References

[1] Hanf JH, Dautzenberg K. A theoretical framework of chain management. Journal on Chain and Network Science. 2006;**6**:79-94

[2] Saitone TL, Sexton RJ. Agri-food supply chain: Evolution and performance with conflicting consumer and societal demands. European Review of Agricultural Economics. 2017; **44**(4):634-657

[3] Miatton F, Amado L. Fairness, transparency and traceability in the coffee value chain through blockchain innovation. In: 2020 International Conference on Technology and Entrepreneurship - Virtual (ICTE-V), San Jose, CA, USA. 2020. pp. 1-6. Available from: https://www.semanticscholar.org/paper/Fairness%2C-Transparency-and-Traceability-in-the-Miatton-Amado/a32812250068dbdb7bca0232d2f26e6ea368f709

[4] Baralla G, Pinna A, Tonelli R, Marchesi M, Ibba S. Ensuring transparency and tracea-bility of food local products: A blockchain application to a Smart Tourism Region. Concurrency and Computation: Practice and Experience. 2020;**33**:e5857. DOI: 10.1002/cpe.5857

[5] Costa C, Antonucci F, Pallottino F, Aguzzi J, Sarriá D, Menesatti P. A review on agri-food supply chain traceability by means of RFID technology. Food and Bioprocess Technology. 2013;**6**:353-366. DOI: 10.1007/s11947-012-0958-7

[6] Katina PF. Systems theory-based construct for identifying metasystem pathologies for complex system governance [PhD], Virginia, USA: Old Dominion University; 2015

[7] Allen D, Berg C. Blockchain governance: what we can learn from the economics of corporate governance (January 15, 2020). In: Allen DWE, Berg C, editors. 'Blockchain Governance: What can we Learn from the Economics of Corporate Governance?', The Journal of the British Blockchain Association (Forthcoming). Available at SSRN: https://ssrn.com/abstract=3519564 or DOI: 10.2139/ssrn.3519564

[8] Kramer MP, Bitsch L, Hanf JH. Blockchain and its impacts on agri-food supply chain network management. Sustainability. 2021;**13**(4):2168

[9] Kramer MP, Bitsch L, Hanf JH. The impact of instrumental stakeholder management on blockchain technology adoption behavior in agri-food supply chains. Journal of Risk and Financial Management. 2021;**14**(12):598. DOI: 10.3390/jrfm14120598

[10] Fishbein M, Ajzen I. Predicting and Changing Behavior: The Reasoned Action Approach. New York: Psychology Press; 2010

[11] Hanf JH. Supply chain networks: analysis based on strategic management theories and institu-tional economics. In: IAMO – Forum 2005. How effective is the invisible hand? Agricultural and Food Markets in Central and Eastern Europe, 16 – 18 June 2005. Conference Proceedings, Halle (Saale), Germany; 2005

[12] Jarillo JC. On strategic networks. Strategic Management Journal. 1988;**9**:31-41

[13] Gulati R, Nohria N, Zaheer A. Strategic Networks. Strategic Management Journal. 2000;**21**(3): 203-215

[14] Gulati R, Lawrence PR, Puranam P. Adaptation in vertical relationships: Beyond incentive conflict. Strategic Management Journal. 2005;**26**:415-440. [CrossRef]

[15] Puranam P and Raveendran M. Interdependence and organization design. In Handbook of Economic Organization. Cheltenham, UK: Edward Elgar Publishing 2013. DOI: 10.4337/9781849803984.00020

[16] Hanf J, Kühl R. Strategy focussed supply chain networks. In: Dynamics in Chain and Networks. Netherlands: Wageningen Academic Publishers; 2004. pp. 104-110

[17] Belaya V, Hanf J. Managing Russian agri-food supply chain networks with power. Journal on Chain and Network Science. 2012;**12**:215-230

[18] Handayati Y, Simatupang TM, Perdana T. Agri-food supply chain coordination: The state-of-the-art and recent developments. Logistics Research. 2015;**8**:1-15

[19] Tripoli M, Schmidhuber J. Emerging Opportunities for the Application of Blockchain in the Agri-Food Industry; FAO: Rome, Italy; ICTSD: Geneva, Switzerland. 2018

[20] Ketchen DJ Jr, Hult GTM. Toward greater integration of insights from organization theory and supply chain management. Journal of Operations Management. 2007;**25**:455-458

[21] Hanf JH, Gagalyuk T. Integration of small farmers into value chains: Evidence from Eastern Europe and Central Asia. In: Agricultural Value Chain. Rijeka: IntechOpen; 2018. DOI: 10.5772/intechopen.73191

[22] Belaya V, Hanf J. Managing Russian agri-food supply chain networks with power. Journal on Chain and Network Science. 2012;**12**:215-230

[23] Giannoccaro I. Centralized vs. decentralized supply chains: The importance of decision maker's cognitive ability and resistance to change. Industrial Marketing Management. 2018;**73**:59-69

[24] Catalini C, Gans JS. Some Simple Economics of the Blockchain (April 20, 2019). Rotman School of Management Working Paper No. 2874598, MIT Sloan Research Paper No. 5191-16. [online] SSRN Electronic Journal; 2019. Available at SSRN: https://ssrn.com/abstract=2874598 or DOI: 10.2139/ssrn.2874598

[25] Buterin, Vitalik. Proof of Stake: How I Learned to Love Weak Subjectivity. Ethereum Foundation Blog, Internet Resource. 2014. Available from: https://blog.ethereum.org/2014/11/25/proof-stake-learned-love-weak-subjectivity/ [Accessed on: 13 June, 2021]

[26] Seebacher S, Schüritz R. Blockchain technology as an enabler of service systems: A Structured Literature Review. In: 8th International Conference on Exploring Service Science. Karlsruhe: IESS; 2017. p. 1.7. DOI: 10.1007/978-3-319-56925-3_2

[27] Quiniou M. Blockchain: The advent of disintermediation. In: Innovation, Entrepreneurship and Management. Hoboken, NJ, USA: John Wiley & Sons; 2019

[28] Aste T, Tasca P, Di Matteo T. Blockchain technologies: The foreseeable impact on society and industry. Computer. 2017;**50**:18-28

[29] Freeman RE. Stakeholder theory. In: Cooper CL, editor. Wiley Encyclopedia of Management. [online] Wiley; 2015. DOI: 10.1002/9781118785317.weom020179

[30] Rowley TJ. Moving beyond dyadic ties: A network theory of stakeholder influences. The Academy of Management Review. 1997;**22**(4): 887-910

[31] Lazzarini SG, Chaddad FR, Cook ML. Integrating supply chain and

network analyses: The study of netchains. Journal on Chain and Network Science. 2001;**1**(1):7-22

[32] Lazzarini SG, Boehe D, Pongeluppe LS, Cook ML. From instrumental to normative relational strategies: A study of open buyer-supplier relations. Vol. 2020, No. 1. Proceedings. Briarcliff Manor, NY, USA: Academy of Management; 2020. DOI: 10.5465/AMBPP.2020.122

[33] Donaldson T, Preston L. The stakeholder theory of the modern corporation: Concepts, evidence and implications. Academy of Management Review. 1995;**20**(1):65-91

[34] Jones TM. Instrumental stakeholder theory: A synthesis of ethics and economics. Academy of Management Review. 1995;**20**:404-437

[35] Mac Clay P, Feeney R. Analyzing agribusiness value chains: A literature review. International Food and Agribusiness Management Review. 2019;**22**(1):31-46. DOI: 10.22434/IFAMR2018.0089

[36] Jones TM, Harrison JS, Felps W. How applying instrumental stakeholder theory can provide sustainable competitive advantage. Academy of Management Review. 2018;**43**(3):371-391. DOI: 10.5465/amr.2016.0111

[37] Hart O. Incomplete contracts and control. American Economic Review. 2017;**107**(7):1731-1752. DOI: 10.1257/aer.107.7.1731

[38] Doh JP, Quigley NR. Responsible leadership and stakeholder management: Influence pathways and organizational outcomes. Academy of Management Perspectives. 2014;**28**:255-274. DOI: 10.5465/amp.2014.0013

[39] Walley P. Stakeholder management: The sociodynamic approach. International Journal of Managing Projects in Business. 2013;**6**(3):485-504. DOI: 10.1108/IJMPB-10-2011-0066

[40] Kramer MP, Bitsch L, Hanf JH. Centralization vs. decentralization - supply chain networks and blockchain in the agri-food business. In: Theory to Practice as a Cognitive, Educational and Social Challenge. Conference Proceedings, Mitrovica, Kosovo: South East European University, Skopje, North Macedonia, International Business College Mitrovica (IBC-M), Kosovo; 2020

[41] Tarhini A, Arachchilage NAG, Masa'deh R, Abbasi MS. A Critical review of theories and models of technology adoption and acceptance in information system research. International Journal of Technology Diffusion (IJTD). 2015;**6**(4):58-77. DOI: 10.4018/IJTD.2015100104

[42] Venkatesh V, Morris MG, Davis GB, Davis FD. User acceptance of information technology: Toward a unified view. Management Information Systems Quarterly. 2003;**27**(3):425-478

[43] Ajzen I. The theory of planned behavior. Organizational Behavior and Human Decision Processes. 1991;**50**(2):179-211. DOI: 10.1016/0749-5978(91)90020-T

[44] Pelupessy W, Díaz R. Upgrading of lowland coffee in Central America. Agribusiness. 2008;**24**:119-140. DOI: 10.1002/agr.20150

[45] Solino. Der Erste seiner Art. 2021. Available at: https://www.solino-coffee.com/ [Accessed on 1 June 2021]

[46] Kaffeekoop. Kaffee ganz aus Frauenhand. 2021. Available at: https://kaffee-kooperative.de/angeliques-finest/ [Accessed: 1 June 2021]

[47] James HS Jr, Sykuta ME. Property right and organizational characteristics of producer-owned firms and organizational trust. Annals of Public

and Cooperative Economics. 2005;**76**:545-580. DOI: 10.1111/j. 1370-4788.2005.00289.x

[48] Wang G, Lu Q, Capareda SC. Social network and extension service in farmers' agricultural technology adoption efficiency. PLoS One. 2020;**15**(7):e0235927. DOI: 10.1371/journal.pone.0235927

[49] Baskaran H, Yussof S, Rahim FA, Abu Bakar A. Blockchain and the Personal Data Protection Act 2010 (PDPA). In: Malaysia, Information Technology and Multimedia (ICIMU) 8th International Conference. Selangor, Malaysia: IEEE; 2020. pp. 189-193

[50] Davidson S, De Filippi P, Potts J. Blockchains and the economic institutions of capitalism. Journal of Institutional Economics. 2018;**14**(4): 639-658

[51] Williamson OE. The Economic Institutions of Capitalism: Firms, Markets, Relational Contracting. New York, NY, USA: The Free Press, A Division of McMillan, Inc.; 1985

[52] Akansel I. Technology in new institutional economics—comparison of transaction costs in Schumpeter's capitalist development ideology. The China Business Review. 2016;**15**:64-93

[53] Pietrewicz L. Blockchain: A coordination mechanism. ENTRENOVA - ENTerprise REsearch InNOVAtion. 2019;**5**(1):105-111

[54] McPhee C, Ljutic A. Editorial: Blockchain. Technology Innovation and Management Review. 2017;**7**:3-5

[55] Kamilaris A, Fontsa A, Prenafeta-Boldú FX. The rise of blockchain technology in agriculture and food supply chains. Trends in Food Science and Technology. 2019; **91**:640-652

[56] Kamble S, Gunasekaran A, Arha H. Understanding the Blockchain technology adoption in supply chains-indian context. International Journal of Production Research. 2018;**57**:2009-2033

[57] Kõlvart A, Fontsa A, Prenafeta-Boldú FX. The rise of blockchain technology in agriculture and food supply chains. Trends in Food Science and Technology. 2019; **91**:640-652

[58] Frantz CK, Nowostawski M. From institutions to code: Towards automated generation of smart contracts. In: Proceedings of the 2016 IEEE 1st International Workshops on Foundations and Applications of Self Systems (FAS*W), Augsburg, Germany, 12-16 September 2016. Augsburg, Germany: IEEE; 2016. pp. 210-215

[59] Pawczuk L, Massey R, Holdowsky J. Deloitte 2019 global blockchain survey - blockchain gets down to business. Deloitte Insights [online]. 2019. Available from: www2.deloitte.com/content/dam/Deloitte/se/Documents/risk/DI_2019-global-blockchain-survey.pdf

[60] Welge MK, Eulerich M. Corporate-Governance-Management. Theorie und Praxis der guten Unternehmensführung, 2. Auflage. Wiesbaden, Germany: Springer Gabler; 2014

[61] Simon HA. Models of Man: Social and Rational. New York: John Wiley and Sons, Inc.; 1957. p. 279

[62] Giannoccaro I. Centralized vs. decentralized supply chains: The importance of decision maker's cognitive ability and resistance to change. Industrial Marketing Management. 2018;**73**:59-69

[63] Simon HA. Models of Man: Social and Rational. New York: John Wiley and Sons, Inc.; 1957. p. 279

[64] Ghosh A, Fedorowicz J. Governance mechanisms for E-collaboration. Encyclopedia of E-Collaboration. 2008:919-925. DOI: 10.4018/978-1-59904-000-4.ch049

[65] Williamson OE. Transaction cost economics. In: Schmalensee RL, Willig RD, editors. Handbook of Industrial Organization. Vol. 1. Amsterdam: Elsevier; 1989. pp. 135-182

[66] Azim MI. Corporate governance mechanisms and their impact on company performance: A structural equation model analysis. Australian Journal of Management. 2012;**37**(3): 481-505. DOI: 10.1177/03128962 12451032

[67] Gulati R, Wohlgezogen F, Zhelyazkov P. The two facets of collaboration: Cooperation and coordination in strategic alliances. The Academy of Management Annals. 2012;**6**(1):531-583

[68] Karame GO, Androulaki E, Roeschlin M, Gervais A, Čapkun S. Misbehavior in bitcoin: A study of double-spending and accountability. ACM Transactions on Information and System Security. 2015;**18, 1**:2. DOI: 10.1145/2732196

[69] Beck R, Müller-Bloch C, Leslie King J. Governance in the blockchain economy: A framework and research agenda. Journal of the Association for Information Systems. 2018; **19**(10):1-35

[70] Lumineau F, Wang W, Schilke O. Blockchain governance — A new way of organizing collaborations? (March 28, 2020). In: Lumineau F, Wang W, Schilke O, editors. "Blockchain Governance—A New Way of Organizing Collaborations?" Organization Science (Forthcoming). Available at SSRN: https://ssrn.com/abstract=3562941

[71] Douma S, Schreuder H. Economic Approaches to Organisations. Harlow, UK: Pearson Education Ltd.; 2013. p. 364

[72] Bitsch L, Atoyan S, Richter B, Hanf J, Gagalyuk T. Including smallholders with vertical coordination. Agricultural Economics. Rijeka: IntechOpen; 2020, DOI: 10.5772/ intechopen.92395

[73] Pass CL, Davies L, Lowes B. Collins Dictionary of Economics. Glasgow: HarperCollins; 2005

[74] Katina P, Keating CB, Sisti JA, Gherorghe AV. Blockchain governance. International Journal of Critical Infrastructures. Print ISSN: 1475-3219 Online ISSN: 1741-8038. 2019;**15**:121-135. DOI: 10.1504/IJCIS.2019.098835

[75] Barki H, Hartwick J. Measuring user participation, user involvement, and user attitude. Management Information Systems Quarterly. 1994;**18**(1):59-82. DOI: 10.2307/249610

[76] Al-Jabri IM, Roztocki N. Adoption and use of information technology in mandatory settings: Preliminary insights from Saudi Arabia (August 12, 2010). In: 16th Americas Conference on Information Systems, August 2010. SSRN; 2010. Available at SSRN: https://ssrn.com/abstract=1965269

Tea Value Chains Viability in Limpopo Province of South Africa: A Cost-Benefit Analysis

Azwihangwisi E. Nesamvuni, James Bokosi,
Khathutshelo A. Tshikolomo, Ndivhudzannyi S. Mpandeli
and Cebisa Nesamvuni

Abstract

The research was conducted to investigate the production of value-added tea as part of the resuscitation of Tshivhase-Mukumbani Tea Estate. Data were mainly obtained from records kept at the Tshivhase-Mukumbani Tea Estate, through a review of literature and interviews of the selected respondents. Evaluation of economic viability of the value-adding initiative was based on Net Present Value (NPC) and the Benefit-Cost Ratio (BCR) calculated from time-series data obtained for the period 2005–2012. The quantity of value-added tea produced varied across years, geographical locations, and seasons, with production higher for wetter seasons. The NPV was consistently negative, while the BCR was below unity throughout the study period, implying that the value-adding initiative was economically not feasible. Initiatives for achieving economic sustainability of the value addition were (1) Improve the marketing of the made tea brand Midi Tea as organic and longer shelf life. (2) Good labor contracting management practices to deal with labor disputes and unrest. (3) Good supply chain and procurement management practices to reduce the cost of production (4) Monitoring the impact of climate variability and mitigate by providing irrigation (5) Intercropping tea with a suitable winter yielding crops such as avocadoes or Macadamia.

Keywords: value-adding, economic viability, Tshivhase-Mukumbani tea estate, benefit–cost ratio, net present value

1. Introduction

The historical background of tea production in South Africa has been in the World of Tea [1]. The narrative indicates that the first time tea was grown was in Durban in the year 1850. The material used for the first planting was imported from London's Kew Gardens. It was only 27 years later that commercialization was emphasized. With the experience of the growers, production was made steady only after the 1960s [1]. Market-oriented tea production in South Africa started in 1964 and the industry was supported by the government and later Industrial Development Corporation (IDC) as one of the "mass employers in the rural areas" and the strategy worked with absorption rates of over 1000 workers on a 500-hectare farm. The effect of democracy in 1994, globalization, and SADC Trade Protocols on agribusiness enterprises was massive and left many agribusinesses battling to remain profitable [1, 2].

The main challenges were high worker minimum wage rates; no protection against tea imports from SADC; high production cost structure (*due to labor wages, electricity, and inputs costs*); the strong brand against the US Dollar; and land claims on tea estates [3].

Value addition and integration of tea production were then adopted by the Limpopo Department of Agriculture in 2006 as a strategy to revitalize the Limpopo Tea Estates [3]. It was an understanding that the extent of the economic viability of the tea business enterprise was going to be influenced by activities performed along the value chain [4]. The thinking has been that by introducing value-adding, business enterprises would be able to increase the revenue and overall profitability of their products [5, 6]. Based on the studies conducted [7–9] indicated that agricultural value-adding initiatives have been recognized as a way of assisting business enterprises to understand the shocks brought about by globalization. Factors influencing commoditization of agricultural products include increasing consumer demand for convenient, ready to eat, safe, and nutritious food products and willingness to pay premium prices for such value-added products [10, 11].

The Tshivhase-Mukumbani Tea Estate adopted the concept of integrated value addition in 2009. This was given impetus by the building of infrastructure inclusive of the erecting of the tea processing plant. Black tea was locally processed, branded, and packaged resulting in the production of the Midi Tea range comprising Midi Gold Tagged, Midi Tag-less tea bags, and Midi loose tea. The focus of this study was to investigate: (a) to establish the production trends prevailing at Tshivhase-Mukumbani Tea Estates under rainfed conditions, (b) the extent to which value-adding implemented by Tshivhase-Mukumbani Tea Estate was economically viable, and (c) considerations for improving the economic viability of the value-adding initiative.

2. Methodology

2.1 Study area

2.1.1 Location

The Tshivhase-Mukumbani Tea Estate is located in the Thulamela Municipality of Vhembe District under Limpopo Province of South Africa (**Figure 1**). The tea estate is comprised of the Tshivhase Farm located at 30.314: 30.367 E and − 22.968: −22.994 S and the Mukumbani Farm located at 30.386: 30.437 E and − 22.904: −22.940 S. The road (Route R523) connecting Thohoyandou with Makhado Town cuts through the Tshivhase Farm with a major portion of the farm on the south of the road while Mukumbani Farm lies on the north of this road.

The road serves as strategic transportation infrastructure for the Tshivhase-Mukumbani Tea Estate. Although tea production at the Tshivhase Farm is rainfed, this portion of the estate lies adjacent to the Vondo Dam, a strategic water reservoir in the Mutshindudi River that supplies water to Thohoyandou and neighboring areas. Production at the Mukumbani Farm is dependent on some supplementary irrigation from the small dam located on the west of this portion of the estate.

2.1.2 Climate

2.1.2.1 Rainfall

Rainfall and temperature are important climatic factors for optimum production of all crops, including tea. Annual rainfall of 2500–3000 mm is considered optimum

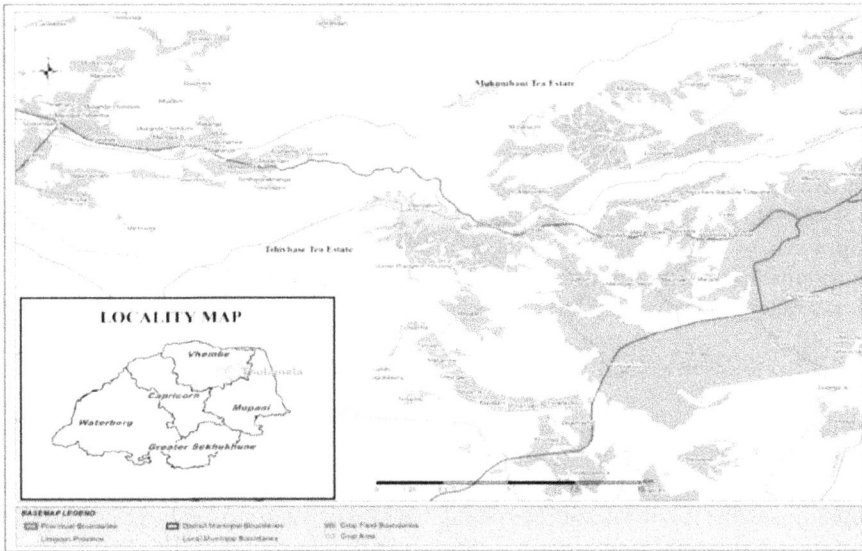

Figure 1.
Location of Tshivhase-Mukumbani tea estate in Vhembe District under Limpopo Province of South Africa (source: GIS unit, Limpopo Department of Agriculture and Rural Development).

for tea production, and the minimum requirement is estimated at 1200 mm [12]. Other than total rainfall received, the distribution of the rainfall is an important determinant of the optimum yield of tea. The annual rainfall at Tshivhase-Mukumbani Tea Estate was 1758.0 mm, well above the minimum requirement for tea production. The distribution of the rainfall was rather uneven (**Table 1**). The Tshivhase- Mukumbani Tea Estate is situated in a micro-climatic area.

Month	Mean monthly rainfall (mm)	Mean monthly rainfall (% of annual)	Mean quarterly rainfall (mm)	Mean quarterly rainfall (% of annual)
January	340.0	19.3	898.0	51.1
February	327.0	18.6		
March	231.0	13.1		
April	90.0	5.1	157.0	8.9
May	41.0	2.3		
June	26.0	1.5		
July	31.0	1.8	126.0	7.2
August	31.0	1.8		
September	64.0	3.6		
October	120.0	6.8	577.0	32.8
November	193.0	11.0		
December	264.0	15.0		
Annual	**1758.0**	**100.0**	**1758.0**	**100.0**

Source: Climate records, Tshivhase-Mukumbani tea estate.

Table 1.
Rainfall distribution at Tshivhase-Mukumbani tea Estate in Vhembe District under Limpopo Province of South Africa.

The rainfall distribution patterns in this area are above normal, in other words, this area receives more than 1750 mm per annum [13]. The rainfall records concurred with the studies conducted by the Agricultural Research Council [13] in 2006. It was noted that half (51.1%) of the rainfall at Tshivhase-Mukumbani Tea Estate was received in the first quarter (January to March) of the calendar year. This followed one-third (32.8%) of the rainfall received in the preceding quarter (fourth quarter, October to December). The second (April to June) and third (July to September) quarters of the year shared the remaining 16.1% of the rainfall. Improvement of soil moisture for increased tea production at the Tshivhase-Mukumbani Tea Estate may be achieved through introduction (Tshivhase Farm) or improvement (Mukumbani Farm) of irrigation.

2.1.2.2 Temperature

The ideal temperature for tea production is 18–25°C with a minimum recommended temperature of 13°C (average for the coldest month) and a maximum of 30°C (average for the warmest month) [12]. Temperatures at Tshivhase-Mukumbani Tea Estate are lower in winter, and these result in tea plants becoming dormant. Excessively high summer temperature may result in wilting of tea leaves, especially at the Tshivhase Farm without irrigation. Not much can be done to change the temperatures for tea production at the estate.

2.2 Research approach

The mixed methods research approach was chosen for its flexibility to combine the attributes of both quantitative and qualitative methods [14]. The multi-methods procedures have the advantage of enlarging the choice of and expanding the investigator's understanding of any study. Researchers [15, 16] defined a quantitative approach as an inquiry into a social or human problem based on testing a theory made up of variables, measured with numbers, and analyzed using statistical procedures to determine whether the predictive generalizations of the theory hold true. On the contrary, the qualitative approach was referred to as an inquiry process of comprehending a social or human problem or phenomenon based on building a complex holistic picture formed with words. It also helps to capture detailed views of informants, which are collected in a local familiar situation [15, 17].

2.2.1 Sampling and data collection

The study employed purposive sampling to select the individuals who had deeper knowledge about overall management and the introduction of value-adding in the tea estate. In that regard, top managers of the tea estate were selected for the study, and those included: the General Manager, Farm Managers, Manager of Value Adding Facility (tea factory), Financial Manager, and Marketing Manager. The selection of these managers was important for both expert and local knowledge to be sourced as this was necessary to respond appropriately to the challenges experienced by the tea estate [18, 19].

As guided by the study on improved competitiveness of the tea industry in South Africa [20], data was mainly obtained from records kept at the Tshivhase-Mukumbani Tea Estate, through a review of literature and interviews of the selected respondents. The literature review focused mainly on scientific journals and books, while interviews focused on the selected managers of the tea estate. Both quantitative and qualitative data were collected, hence, the research method was described as mixed [16, 21].

2.2.2 Analytical technique

To determine the status regarding the economic viability of the value addition initiative, the Net Present Value (NPV) and Benefit–Cost Ratio (BCR) were used. An initiative or project is considered financially viable when the NPV is positive and the BCR is greater than unity [22]. To determine the NPV for the value-added Mukumbani-Tshivhase Tea Estate, the cash flow was determined by deducting the Cash Outflow from the Cash Inflow to obtain the Net Cash Flow (NCF), and that (NCF) was discounted. Discounting may be described as the opposite of compounding [23]. The standard models for determining NPV and BCR are as follows:

1. The Net Present Value (NPV)

$$NPV = \sum_{t=0}^{n} (Bt - Ct) / ((1+i)^{\wedge} t)$$

2. The Benefit–Cost Ratio (BCR)

$$BCR = \sum_{t=0}^{n} (Bt / (1+i)^{\wedge} t) / (Ct / (1+i)^{\wedge} t)$$

Where:
B_t = Benefits in year t.
C_t = Costs in year t.
n = Number of years the initiative or project will be under operation.
i = Discount rate.

3. Results and discussion

3.1 Tea production trends at Tshivhase-Mukumbani since revitalization in 2006

The Tshivhase tea project was projected to be sustainable under a fully integrated strategy where the enterprise was able to package black tea, market, and distribute it for retail and wholesale market outlets. The diagrams in this section seek to depict the impact of yearly and seasonal variation since the rehabilitation of the tea estates in 2006.

3.1.1 Annual trends in tea production parameters

3.1.1.1 Trend in green leaf production (kg) in Tshivhase and Mukumbani estates

Figure 2 shows the trend in green leaf production (Kg) in Tshivhase and Mukumbani estates from 2007 to 2013. Tshivhase estate (TTP) is bigger in size of 577 ha compared to Mukumbani which is 500 ha. The decrease in the production of green leaf in 2011–2012 was due to an industrial strike that started in September 2011 and ended in June 2012. Production of the green leaf then picked up slowly from October 2012 onwards. The size of the estate in a large part explains the large differences of 371,729 kg in 2007–2008 financial year to 827,803 kg in 2010–2011 financial year green leaf production between the two estates. Within the industry domain, there has been a school of thought to the effect that competitiveness is higher in estates compared to smallholdings.

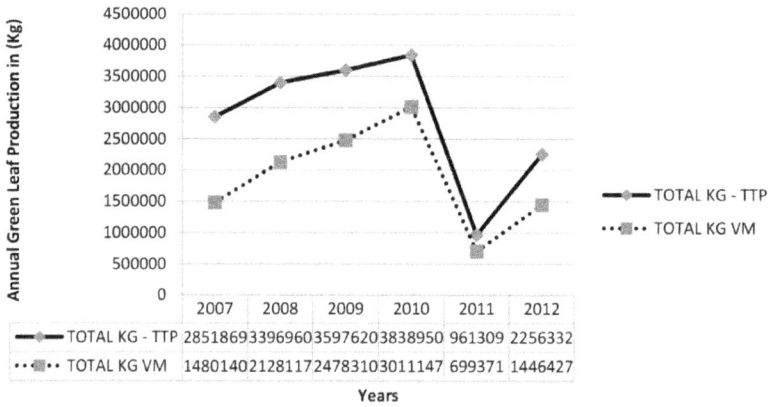

Figure 2.
Annual green leaf production in (kg) in Tshivhase (TTP) and Mukumbani (VM) estates from the year 2007 to 2012.

The challenge with small farms is that they do not have the advantage of the processing profits of made tea that accrue to big estates as small farm's activities end in the harvesting of green leaves. The second school of thought is to the contrary, advocating for small farms on the basis that there tends to be a higher output per workday and higher land competitiveness per unit of area. For tea to be profitable efforts should be to ensure a model that led to processing and or any value-adding especially where smallholder farmers are involved. The model should be coupled with a strong and efficient advisory system with appropriate factory infrastructure for processing the green leaves [24–27].

3.1.1.2 Trends man-days in Tshivhase (TTP) and Mukumbani (VM) estates

Central to the production of green leaf is the demand for labor to do the plucking for processing. **Figure 3** indicates the annual estimated labor per unit in Tshivhase (TTP) and Mukumbani (VM) estates from the year 2007 to 2013. The trend follows the same pattern as that of **Figure 2**, mainly because these are the number of days that a labor force equivalent to 12 people per hector is deployed to pluck the tea, weed, clear the bushes table, and all other labor-related activities. The labor force in Tshivhase is always higher than that of Mukumbani which also relates to the size of the estate as indicated in Section 3.1.1.1. There seems to be a consensus that the workday requirements and the labor per unit of area is the central measure on which to improve the competitiveness of tea enterprises.

It should be indicated that these measures of efficiency will vary from one nation to the other, region and continents. Technologies being introduced to the industry also create a variation worth registering through the introduction of computer programming [27].

3.1.1.3 Trends in green leaf production in (kg) per labor unit in Tshivhase (TTP) and Mukumbani (VM) estates

The trend indicated in **Figure 4** is the amount of green leaf produced per day per worker. Tshivhase Tea Estate had higher production per labor unit as follows: 10.16 in 2007–2008, 3.82 in 2008–2009, 5.55 in 2009–2010 with very little differences

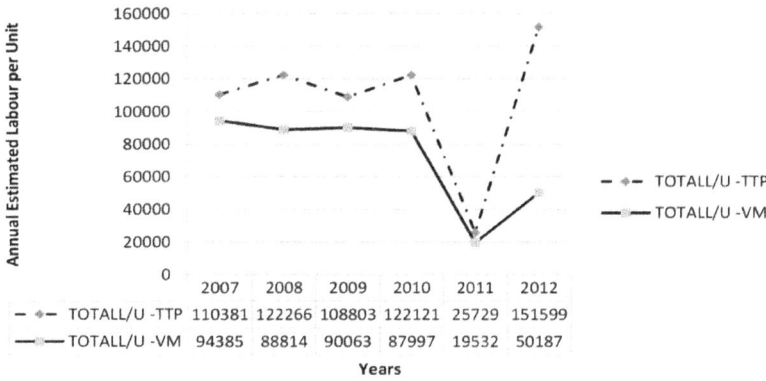

	2007	2008	2009	2010	2011	2012
TOTALL/U -TTP	110381	122266	108803	122121	25729	151599
TOTALL/U -VM	94385	88814	90063	87997	19532	50187

Figure 3.
Annual estimated man-days in Tshivhase (TTP) and Mukumbani (VM) estates from the year 2007 to 2012.

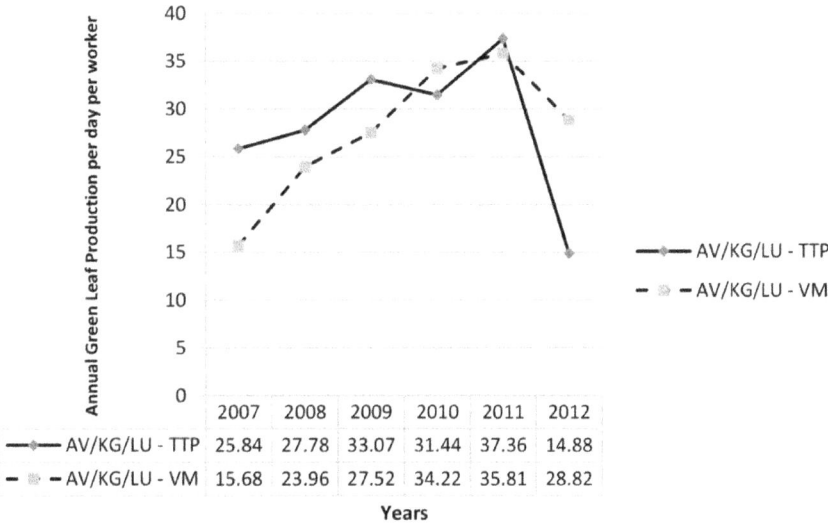

	2007	2008	2009	2010	2011	2012
AV/KG/LU - TTP	25.84	27.78	33.07	31.44	37.36	14.88
AV/KG/LU - VM	15.68	23.96	27.52	34.22	35.81	28.82

Figure 4.
Annual average green leaf production in (kg) per labor unit in Tshivhase (TTP) and Mukumbani (VM) estates from the year 2007 to 2012.

in the other years except a marked difference in 2012–2013 of 13.94 higher for Mukumbani compared to Tshivhase. The production of the green leaf at Tshivhase and Mukumbani per day per worker was lower compared to other countries. The recorded figures that were also developed in trends over the years show an amount of 36 kg green leaf per day per worker. The average green leaf produced per day per worker was found to be about 24 kg in India and private estates in Sri Lanka. The values were far less when compared to between 40 to 50 kg per day per worker recorded in Kenya. Higher values were recorded in Zimbabwe to as high as 68 kg per day per worker [2, 28].

Due to shortages of labor, there has been an attempt to introduce novel labor schemes and mechanical harvesters especially in periods with high leaf volumes. The biggest disadvantage has been its effect on the leaf quality that will yield extremely poor made tea despite its effect on the savings on labor [26, 27].

3.1.1.4 Trend in minimum wage for both Tshivhase and Mukumbani

Tea production is a land-intensive and a labor-intensive enterprise. Both these factors of production especially labor is now scarce and costly. Labor at Tshivhase and Mukumbani was governed by the South African Sectorial Determination (minimum wage). The minimum wage increased from R885.00 in 2007–2008 financial years to R2274.84 in 2014 (**Figure 5**). The percentage increase was zero in 2007 to 51 percent in 2013 2014 financial year. Subsequently, government intervention on labor led to an increase of 61 percent over a period of 7 years. It is worth noting that workforce requirements should be based on competitiveness levels for existing tasks. For Tshivhase and Mukumbani new cost-effective norms must be established. Due to the increase in the minimum wage, there may be a need to reduce the man-days per labor unit from 12 to eight in order to hire less labor for the activities at hand. When that happens and labor competitiveness improves, there will be savings on labor which, in turn, will have a salutary effect on the cost of production [2, 28, 29].

3.2 Value-added tea production at Tshivhase (TTP) and Mukumbani (VM) estates

3.2.1 Trends in made tea production in (kg) in Tshivhase (TTP) and Mukumbani (VM) estates

Figure 6 shows the production trend of made tea which is green leaf processed into black tea. The highest production figures were in the year 2010–2011 with Tshivhase at 884064 kg and Mukumbani at 625767 kg. This was followed by the years 2008–2009 which recorded the figures of 739,788 at Tshivhase and 534,286 at Mukumbani and 2009 and 2010 years with records of 760,729 at Tshivhase and 525,303 at Mukumbani respectively. The lowest production was in 2006–2007 at the start of the rehabilitation and the 2011–2012 due to the industrial strike. The standards set by the tea industry are that the factory requires at least 4.5 kg of green leaf to process 1 kg of tea. The challenge in many parts of the world is that the land in the amount of made tea that can be produced. The processing methods also create a variation in the amount of made tea from green leaves [25].

Comparison between the CTC (cut, tear, and curl) and the orthodox method shows that the CTC gives more cups of tea and quick brewing than the orthodox method. This has led to the shift in processing methods to CTC at the ratio of 85:15

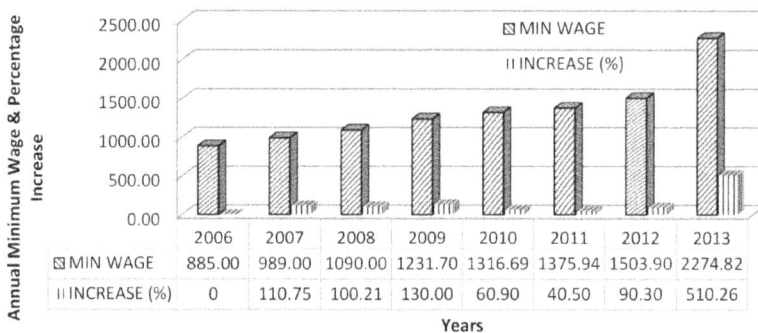

Figure 5.
The annual minimum wage with associated percentage increase factored by 10 from the year 2006 to 2013 for both Tshivhase and Mukumbani.

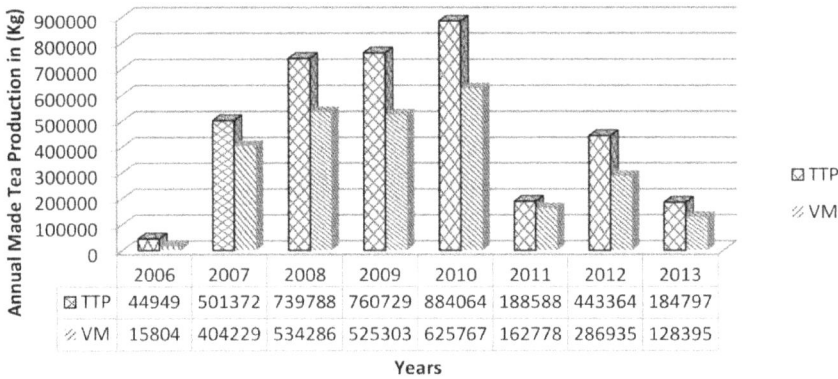

	2006	2007	2008	2009	2010	2011	2012	2013
TTP	44949	501372	739788	760729	884064	188588	443364	184797
VM	15804	404229	534286	525303	625767	162778	286935	128395

Figure 6.
Annual made tea production in (kg) in Tshivhase (TTP) and Mukumbani (VM) estates from the year 2006 to 2013.

in India, 10: 90 in Sri Lanka, and 100 percent in Bangladesh. The competitive advantage of a shift into CTC is that labor requirement is half that of the orthodox method across all types of tea estates [25–27].

Annual quantities of made tea, production was the lowest at 60753 kg in 2006/2007, and this was probably due to the fact that the tea scrubs had to recover from the shock of having been neglected in previous years. The annual production of made tea continually increased up to 2010/2011 cycle with a maximum of 1509 831 kg recorded, followed by a decline in 2011/2012 cycle and some increase in 2012/2013 (**Figure 6**). In the 2011/2012 cycle, the estates experienced major labor unrest which cause a decline in the production of made tea. The experience of the tea estate management (R.W. Topham - personal communication, June 06, 2016), the tea estate had the potential to produce up to 2 million kg of made tea per annum (about 1 million from the Tshivhase Farm and 1 million from the Mukumbani Farm).

3.2.2 Seasonal production in (kg) of made tea in Tshivhase and Mukumbani tea estates

The ultimate income to any tea estate is the amount of made tea after processing. **Figure 7** shows the production of made black tea as influenced by seasonal variations. The pattern on the monthly variations follows the same trend as that of green leaf production. Due to the lag that occurs between when the first leaf is picked to production and the logistic of processing that takes only at the most 2 days the data recorded shifts by almost a month. Made tea is lowest in production in July August and September. Made tea is also highest in December, January, February, and March. The only year where there was minimal black tea made was 2006–2007.

The lowest quantity (11,991 kg) of made tea was in the months of spring (July to September) of the year (**Figure 7**), probably a result of the fact that rainfall was the lowest (126 mm, 7.2% of annual rainfall) during this season (**Table 1**). This had an impact on the yields and ultimate quantity of made tea produced. An increased amount of rainfall (577 mm, 32.8% of annual) was received in the summer season, and this was followed by an increased production (212,880 kg) of made tea. As the rainfall continued to increase (898.0 mm, 51.1% of annual) in the summer of the following year, so was the production of made tea (462,073 kg). The production of made tea fell in the second quarter as the rainfall declined.

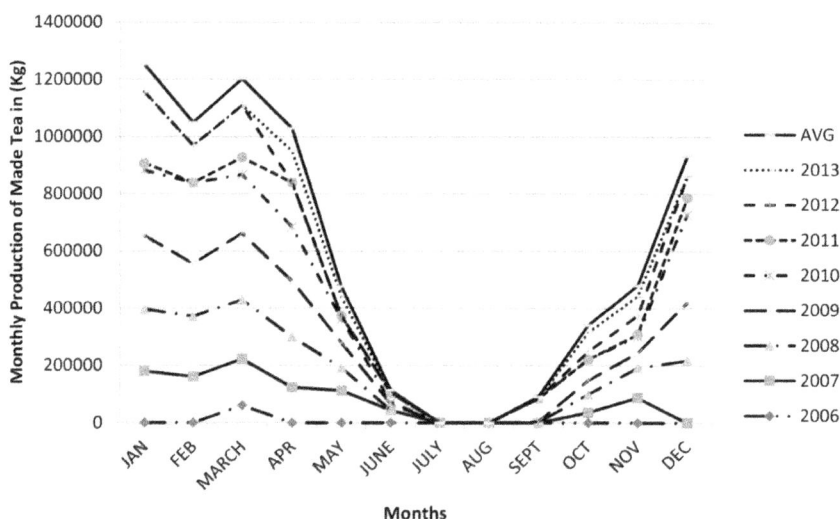

Figure 7.
Monthly production made tea in (kg) from 2006 to 2014 across Tshivhase and Mukumbani tea estates.

3.2.3 Net present value

The NPV is the difference in the present value of cash inflows and the present value of cash outflows and is used in capital budgeting to analyze the profitability of a projected investment (Firer et al., 2012). A project is considered economically viable when the NPV is positive [22, 30]. The NPV for Tshivhase-Mukumbani Tea Estate was consistently negative throughout the study period (**Table 2**). The NPV fluctuated showing a decrease (becoming more negative) from 2005 to 2008 after which it increased (became less negative) to 2010 and again decreased to 2012. For the whole study period, the NPV was highly negative (− R153 530,525–70) and this implied that the value-adding initiative was economically not feasible [22, 31, 32].

3.2.4 Benefit–Cost ratio

The BCR is a ratio attempting to identify the relationship between the cost and benefits of a proposed project, [22]. The BCR for Tshivhase-Mukumbani Tea Estate

Year	Total cost (R)	Revenue (R)	Discount factor (12%)	Net present value
2005	3,416,481.0	46,900.0	0.12	−3,374,606.00
2007	8,275,029.0	100,038.0	0.12	−8,195,279.32
2008	29,764,438.0	382,293.0	0.12	−29,492,329.39
2009	36,856,647.0	18,946,979.0	0.12	−24,815,499.31
2010	22,777,283.0	20,053,808.0	0.12	−11,398,213.78
2011	46,401,574.0	19,145,523.0	0.12	−36,701,856.22
2012	48,561,787.0	19,916,129.0	0.12	−39,552,741.67
Total	**196,053,239.0**	**78,591,670.0**		**−153,530,525.7**

Source: Financial records, Tshivhase-Mukumbani tea estate.

Table 2.
Net present value for Tshivhase-Mukumbani tea estate for the period 2005 to 2012.

was less than unity throughout the study period (**Table 3**). For the period 2005 to 2008, the BCR values recorded were much less (BCR <0.05), suggesting that much less benefits were derived for the value-adding investment during this period. The situation improved for the period 2009 to 2012 (0.4 < BCR < 1) although the BCR remained less than unity. The fact that the BCR was less than unity implies that the value-adding activity at Tshivhase-Mukumbani Tea Estate was economically unfeasible. The result of the BCR analysis supports that of the NPV analysis, and this confirms that the value-adding initiative at the tea estate understudy was economically unfeasible. The findings from the economic analyses suggest that the Tshivhase-Mukumbani Tea Estate should consider revising its strategies if it wants the investment in value-adding initiatives to be economically feasible.

3.3 Essential strategies for achieving economic feasibility

Considering the model used to determine the NPV and that for BCR, achievement of economic feasibility (a positive NPV and BCR > 1) requires an increase of revenue and/or a decline of total costs. Revenue is a product of *Quantity of produce* and *Unit price* while total costs are comprised of *establishment, fixed, and variable costs.*

3.3.1 Quantity of tea produced

The quantity of tea produced at the Tshivhase-Mukumbani Tea Estate was highly influenced by such issues as moisture availability and production management.

3.3.1.1 Increase supply of soil moisture

The quantity of processed tea produced at Tshivhase-Mukumbani Tea Estate was highly influenced by rainfall and its influence on moisture availability. The rainfall in the area was less than the 2500 mm-3000 mm required for optimum production [12], and this resulted in a reduced quantity of tea produced. Increasing the supply of soil moisture through irrigation would therefore result in a rising yield of tea leaves and subsequently increased quantity of value-added tea. Such an irrigation initiative could be introduced at the Tshivhase Farm while more water could be availed for the current supplementary irrigation at Mukumbani Farm. The initiative of increasing soil moisture supply through irrigation would need to be considered for the warm months of the year when temperatures are suitable for tea growth. In

Year	Total cost	Revenue	Discount factor (12%)	BCR
2005	3,416,481.0	46,900.0	0.12	0.013727575
2007	8,275,029.0	100,038.0	0.12	0.012089142
2008	29,764,438.0	382,293.0	0.12	0.012843952
2009	36,856,647.0	18,946,979.0	0.12	0.514072238
2010	22,777,283.0	20,053,808.0	0.12	0.880430208
2011	46,401,574.0	19,145,523.0	0.12	0.412605034
2012	48,561,787.0	19,916,129.0	0.12	0.410119360
Total	196,053,239.0	78,591,670.0	0.12	0.400869021

Source: Financial records, Tshivhase-Mukumbani tea estate.

Table 3.
Benefit–cost ratio for Tshivhase-Mukumbani tea estate for the period 2005 to 2012.

some places in the district tea can only be produced under irrigation sine Vhembe District is known for high climatic variability and change [33].

3.3.1.2 Improve production management

Production of made tea was far less than the potential of the estate, and this was perceived to be partly a result of poor management (R.W. Topham, personal communication, June 06, 2016). Mukumbani-Tshivhase Tea Estate occasionally had some labor unrests resulting in extended work stoppages during the 2011/2012 cycle. Human Resources Management especially labor contracting as part of recruitment is, therefore, necessary to avoid labor disputes and unrest. This in turn will save on time lost during the unrest and increase the quantity of tea produced, both infield and at the factory for the production of made tea.

3.3.2 Increase unit price of sold tea

The unit price of tea produced at the estate was influenced by value-adding and by market demand.

3.3.2.1 Increase sales of value-added tea

The essence of the value-adding initiative at Tshivhase-Mukumbani Tea Estate was to sell the made tea at competitive prices. Although the value-adding initiative was performed well, a lot of the tea was still sold in bulk (only partially value added) at low prices. The tea should be sold as a complete value-added product in order to fetch higher prices, and this is influenced by the market demand for the tea.

3.3.2.2 Increase market demand for value-added tea

The unit price of tea sold by the estate was highly influenced by the market demand for the tea. The value-added tea had not performed well in the market compared to its competitors. As a result, the value-added tea from the estate occupied only a small shelf space in the market, hence a lot of the tea was sold in bulk at low prices. Thorough market research would therefore be necessary, and this should result in improved marketing strategies that increase the market demand for the tea brand (Midi Tea) produced at Tshivhase-Mukumbani Tea Estate.

3.3.3 Reducing costs

Total costs may be calculated as the sum of establishment costs, fixed costs, and variable costs. Establishment costs at Tshivhase-Mukumbani Tea Estate were mostly for the establishment of infrastructure, mainly fencing, irrigation, storage, housing, and value-adding infrastructure. For the period under investigation, the major establishment costs incurred were on constructing and equipping the tea value-adding facility. The cost for the establishment of the value-adding facility was R196 053239–00 (**Table 2**), which was high.

The higher cost may have been due to the practice of inflating costs of construction by contractors without appropriate monitoring by government project managers. As these costs were historic in nature there was no intervention to reduce them. Fixed and variable costs incurred in production and value-adding operations should, where possible, be reduced for the value-adding initiative to be economically feasible. Such costs would include those incurred on labor, maintenance of infrastructure and machinery, electricity, water, and other consumables. Effective cost reduction would require proper management for efficient use of resources. Good labor contracting coupled with supply chain and procurement management practices can optimize the production of made tea.

4. Conclusions

The quantity of rainfed value-added tea produced varied based on year and season of production. Production of both unprocessed and made tea is perceived to be below the potential of the tea estate. The NPV was consistently negative while the BCR was below unity throughout the study period, implying that the value-adding initiative was economically not sustainable. However, the study only covered a period of not more than 7 years. A further study is recommended with outer years covering at least 20 years. The following major initiatives are recommended for achieving economic sustainability of the value addition were: (1) Improve the marketing of the made tea brand Midi Tea as organic and fresh.

The whole production value chain from picking to packaging done in the same estate; (2) Good labor contracting management practices to deal with labor disputes and unrest; (3) Good supply chain and procurement management practices to reduce the cost of production; 4) To increase the quantity of value-added tea produced, it would be necessary to monitor the impact of climate variability and mitigate by providing irrigation especially in the spring and summer months and (5) Tea production at the estates is seasonal, for long term economic sustainability there is a need for inter-cropping tea with a suitable winter yielding crops such as Avocadoes or Macadamia.

Acknowledgements

The authors would like to acknowledge the contribution of the Water Research Commission, South Africa for funding the research theme of the Lead Researcher – Prof AE Nesamvuni.

Conflict of interest

The authors declare no conflict of interest.

Notes/thanks/other declarations

The authors appreciate the support of the University of Free State Executive Management, Dean of Faculty, and Director of the Center for Sustainable Agriculture.

Author details

Azwihangwisi E. Nesamvuni[1*], James Bokosi[1,2], Khathutshelo A. Tshikolomo[1,3], Ndivhudzannyi S. Mpandeli[1,4] and Cebisa Nesamvuni[1,5]

1 Centre for Sustainable Agriculture and Extension, University of the Free State, Bloemfontein, South Africa

2 Gauteng City College, Polokwane, South Africa

3 Limpopo Department of Agriculture and Rural Development, Polokwane, South Africa

4 Water Research Commission of South Africa, South Africa

5 Department of Nutrition, University of Venda, Thohoyandou, South Africa

*Address all correspondence to: nesamvunie@gmail.com

IntechOpen

References

[1] Twinnings. History of South African Tea. 2011. Available from: http://www.twinings.com/enint/worldof tea/southAfricaprof.html

[2] Statistical Bulletin of the Bangladesh Tea Board, 2010, 2020. Available from: http://www.teaboard.gov.bd/sites/default/files/files/teaboard.portal.gov.bd/monthly_report/734d0e2a_ceb7_4e26_82f1_10c2b654365b/2021-01-21-17-56-640c67fbbaacee97a387dd83644463f6.pdf.

[3] LDA. Limpopo Department of Agriculture Report on the Rehabilitation of Tshivhase – Mukumbani Estates. Polokwane: Government of Limpopo;

[4] Harris T. Value-Added Public Relations: The Secret Weapon of Integration. Lincolnwood: NTC Business Books; 1998

[5] Amanor-boadu V. Enhancing competitiveness with value adding business initiatives: Economics and strategy issues. In: Risk and Profit Conference. Kansas State University; 2003

[6] South African Breweries PLC. Corporate accountability report. 2002. Available from: https://www.sharedata.co.za/Data/000125/pdfs/SABMILLER_ar_02.pdf

[7] Ja Afar-Furo M, Bello K, Sulaiman A. Assessment of the Prospects of Value Addition among Small Scale Rural Enterprises in Nigeria: Evidence from North-Eastern Adamawa. Mubi, Bauchi, Nigeria: Adamawa State University, Bauchi State Development Programme, Department of Agricultural Economics and Extension, Department of Planning, Monitoring and Evaluation; 2011

[8] Coltrain D, Barton M, Boland M. Value Added: Opportunities and Strategies. Arthur Capper Cooperative Center, Cooperative Extension Service, Department of Agricultural Economics, Lawrence, Kansas: Kansas State University; 2000

[9] Amanor-boadu V. Trade Liberalization and the World Trade Organisation Negotiations after Seatle. Guelph: George Morris Centre; 2000

[10] ERS/USDA. Animal Products Branch, Economic Research Service. 2002. Available from: http://www.ers.usda.gov/Briefing/FoodPriceSpreads/spreads/table1a.htm [Accessed: 24 May 2012]

[11] Nilson T. Competitive Branding: Winning in the Marketplace with Value-Added Brands. New York: John Wiley & Sons; 1998

[12] Waheed A, Hamid FS, Ahmad H, Aslam S, Ahmad N, Akbaker A. Different climatic data observation and its effect on tea crop. Journal of Materials and Environmental Science. 2013;**4**(2):299-308. ISSN: 2028-2508

[13] Mpandeli NS. Coping with climate variability in Limpopo Province [Unpublished PhD thesis]. South Africa: University of the Witwatersrand, Johannesburg; 2006

[14] Tashakkori A, Teddlie C. Mixed Methodology: Combining Qualitative and Quantitative Approaches. Thousand Oaks: Sage Publications; 1998

[15] Creswell JW. Research Design: Qualitative, Quantitative and Mixed Methods Approaches. 2nd ed. Thousand Oaks: Sage Publications; 2003

[16] Leedy PD, Ormrod JE. Practical Research, Planning and Design. 8th ed. New Jersey: Pearson Merrill Prentice Hall; 2010

[17] Sharique A, Saeeda W, Sumaiya I, Sudarshana G, Anshika S, Zarina F. Qualitative v/s. quantitative research: A summarized review. Journal of Evidence Based Medicine and Healthcare. 2019;**6**(43):2828-2832. DOI: 10.18410/jebmh/2019/587

[18] Hermans LM. Exploring the promise of actor analysis for environmental policy analysis: Lessons from four cases in water resource management. Ecology and Society. 2008;**13**(1):21. [online]. Available from: http://www. ecologyandsociety.org/vol13/iss1/art21/

[19] Tshikolomo KA. Development of a water management decision model for Limpopo Province, South Africa [Unpublished PhD thesis]. Bloemfontein, South Africa: University of the Free State; 2012

[20] Nesamvuni AE, Tshikolomo KA, Nephawe KA, Topham RW, Mpandeli NS. Employee perceptions on determinants of tea competitiveness: A case of Tshivhase-Mukumbani Estate in Limpopo Province of South Africa. International Journal of Agricultural Extension. 2014;**02**(03):193-203. ISSN: 2311-6110

[21] Hurmerinta-Peltomaki L, Nummela N. Mixed methods in international business research: A value-added perspective. Management International Review. 2006;**46**(4):439-459. Available from: www.jstor.org/stable/40836097

[22] Firer C, Ross SA, Westerfield RW, Jordan BD. Fundamentals of Corporate Finance. 5th South African ed. Shoppenhangers Road, Maidenhead, Berkshire: McGraw-Hill Education (UK) Limited; 2012

[23] Correia R, Flynn D, Uliana E. Financial Management. 6th ed. Cape Town: Juta; 2007

[24] Chatterjee DN. Review of ILO/UNFPA/Indian Tea Association Family Welfare Education Project in Dooars. India: Ceylon Labour Foundation; 1982

[25] Sivaram B. Productivity improvement and labour relations in the tea industry in South Asia, ILO. In: 100th Session of International Labour Conference. Geneva: ILO; 1994. Available from http://www.ilo.ch/public/english/dialogue /proasia2.htm

[26] Baffes J. Tanzanian Tea Sector Constraint and Challenges. Washington DC: World Bank; 2003

[27] Melican N. Potential Production of High-Quality Orthodox Tea in Rwanda, and Organizational Constraints at Kitabi and Mata Units. ADAR Project: International Inc; 2004

[28] India and Sri Lanka. Tea Statistics. 2010. Available from: http://www. teaboard.gov.in/pdf/57th_Annual_Report_2010_2011_pdf7318.pdf

[29] Wesumperuma D, Gooneratne W, Fernando NA. Labour absorption in the plantation crop sector of Sri Lanka. 1985. Available from: https://www. srilankateaboard.lk

[30] Mosoma M, Belete AS, Masuku M. The economics of smallholder dairy goat production in Mafefe community of Limpopo Province, South Africa. Asian Journal of Agricultural Sciences. 2012;**4**(2):153-157. ISSN: 2041-3890

[31] Gittinger J. Economic Analysis of Agricultural Projects. 2nd ed. Baltimore: John Hopkins University Press; 1982

[32] The Standard Bank of South Africa Limited. Finance and Farm Management. 5th ed. Johannesburg, Gauteng, South Africa: Standard Bank Agribusiness SA; 2013

[33] Maponya P, Mpandeli NS. Climate change status in Mutale local municipality: A case study of smallholder farmers in Vhembe District, Limpopo Province, South Africa. Journal of Human Ecology. 2015;**52**(12):1-8. DOI: 10.1080/09709274.2015.11906924

Local Tubers Production and Value Chain: Evidence from Sangihe Island Regency, Indonesia

Agustinus N. Kairupan, Gabriel H. Joseph,
Jantje G. Kindangen, Paulus C. Paat, August Polakitan,
Derek Polakitan, Ibrahim Erik Malia and Ronald Hutapea

Abstract

Sangihe Islands Regency is an archipelago located in the border area of the Philippines. The area is far from the Provincial Capital and the Indonesian capital of Jakarta. Therefore, it is sometimes difficult for people to access basic needs, especially food. On the other hand, they have access to alternative food consumed hereditary. For instance, there are plenty of tuber food crops, including cassava, sweet potatoes, and taro. Thus, there is a need for discussion in empowering people on the benefits of tubers such as productions, value chain, and potential development. The methodological approach used is descriptive exploratory, where the data collected is secondary from the desk review related to the potential and food conditions of the people in the area. Several local tuber crops are suitable for development as a staplefood on Sangihe Islands. The development supported by the adequate technological application can optimally increase product value and revenue. Furthermore, those aspects need systematically and synergistically patterned.

Keywords: border area, food security, local tubers, production, technology

1. Introduction

The issue of food security has been developing for a while, both in the international community and in the national community in Indonesia. In several processes and forms of national food security, the government promotes local food-based community development [1, 2]. Alternative ideas for realizing national food security are not only important but should become a massive national movement to guarantee people's quality of life at the local, national and international levels. Food security is not only the problem of producing rice or other crops to be consumed as they are. However, in a broad sense, food security includes how society at local and national levels can produce other crops than rice such as corn, tubers, cassava, sago, and so forth. At the local level, the production depends on the area where the community resides [1, 3].

Communities on the border areas, whose territories are archipelagic areas, are vulnerable to food insecurity. The availability of food on the border with wider natural conditions, and the sea area makes this place highly dependent on food supplies from outside the region. The availability of food in the regions must be optimized to meet food needs. Utilization of the potential local food crops will be important in efforts to meet food needs so that the community will be in a food security position.

Some of the main developing issues are related to food security in border areas, including socio-economic inequality between people living in border areas and neighboring countries, relatively low agricultural productivity, limited information and technology dissemination, inadequate infrastructure, availability of infrastructure and facilities, distribution of land and between islands that can reach all regions. Thus, the inability of the poor to provide sufficient food in terms of nutrition and food security has not become a major concern [4, 5].

An important aspect in achieving food security for people at the border area is the ability to empower local food sources. To anticipate this, it is necessary to bring back local food sources [6]. States that local food has advantages in terms of quality, quantity and also functions for the preservation of biodiversity. The border area of the Sangihe Islands Regency has several types of local food such as tubers which are sources of staple food for the area.

The way to build food self-sufficiency in small islands and border areas is very wide open because the people in the area have been familiar with non-rice food sources for decades and the area does have local food sources that can substitute rice [7]. The increase in food production capacity is carried out based on the potential of agricultural resources. The development of food production is not based on a specific commodity approach, but rather on potential commodities in each region that can be developed into local food sources by increasing production and product processing [8]. Regarding the development of tuber production as local food in border areas, the current condition tends to decrease both in terms of cultivation and utilization of the product due to the declining preference for local food and the lack of intervention in preservation [9]. Local food management in border areas is still limited to traditional techniques and products are not managed with an optimal farming system. To manage local food resources, technological innovation is needed, starting from the cultivation stage to the processing stage to ensure the availability of raw materials for processed food.

Technological innovation for the use of local food needs to be directed at increasing added value, competitiveness, and improving production technology to produce products that are following the wishes and needs of the community (demand-driven) [10]. The increase in added value and product competitiveness is the difference between the potential selling value of the product and the costs required for production. The added value and competitiveness of the product can attract investors to participate in developing the local food agroindustry. Improvements in local food technology, among others, are directed at producing products that are easy and practical to process and consume, with taste and quality that are following market demands. The products should also taste good and be packaged attractively, as well as easy to access (continuity of product availability). The technology developed should be adapted to the needs of the community and the growing market, so that it can compete with other products. Technological improvements will provide opportunities for the realization of product diversity that provides opportunities for consumers to choose products that truly suit their needs and preferences [11, 12].

This paper presents the potential, problems, support for technological innovation, added value, and strategies for developing local tuber food in the border area of the Sangihe Islands Regency.

2. Methodological approach

The present research design is the scientific review method. This method is used to conduct descriptive exploration and data analysis regarding the topics discussed, which are sourced from various scientific references, both from research reports and relevant journals.

3. General conditions of the border area

3.1 Regional characteristics

The Sangihe Islands Regency, located in North Sulawesi Province, is a region in Indonesia that is directly adjacent to the neighboring Philippines (Article VII of Law No. 77 of 1957). This area is the gateway and northern fortress of the Unitary State of the Republic of Indonesia and is a cross-border trade area (Article II of Presidential Decree No. 6 of 1975). In developing the management of islands in border areas throughout Indonesia, where priority is directed to agricultural development planning for certain commodities according to the carrying capacity of the island, the agricultural sector is the main source in meeting basic life needs, especially food, and plays an important role in the economy of the region.

The Sangihe Islands Regency is geographically an integral part of North Sulawesi Province with Tahuna as the capital. It is about 142 nautical miles from the Capital of North Sulawesi Province, Manado, located between 20 4′13″–40 44′22″ North Latitude and 1250 9′28″–1250 56′ 57″ East Longitude. Its boundaries are as follows: North-Republic of the Philippines and the District of the Talaud Islands; South-Sitaro Regency; East—the Pacific Ocean and Maluku Sea; West-North Sulawesi

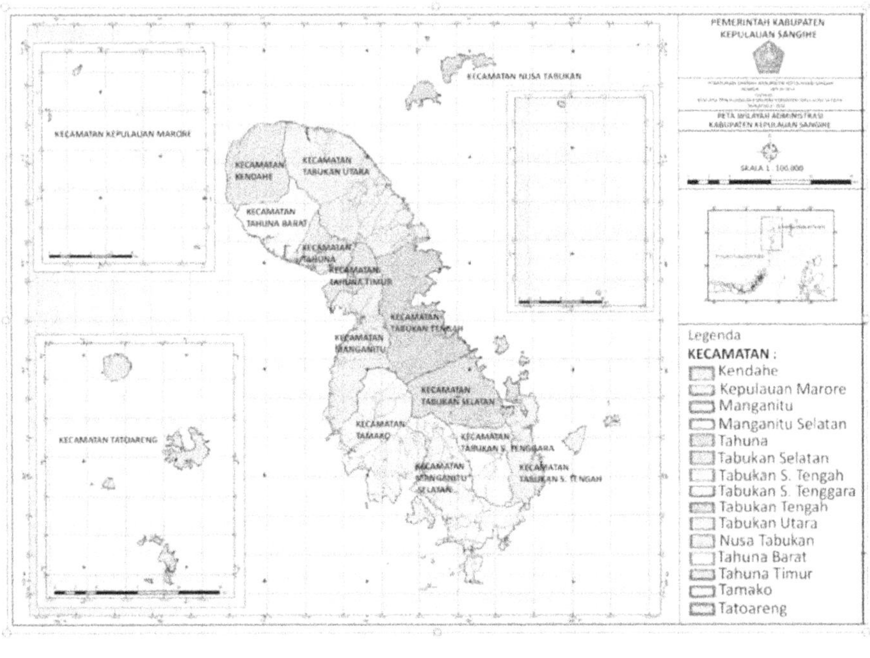

Figure 1.
Regional Map of Sangihe Islands Regency.

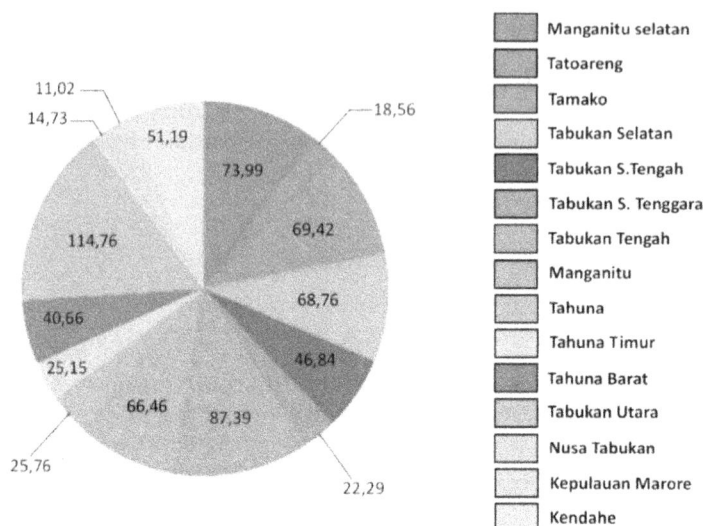

Figure 2.
Sangihe Islands Regency by District year 2016 [13].

(**Figure 1**). The area is 736.98 km^2 divided into 15 sub-districts (**Figure 2**). North Tabukan is the sub-district with the largest area, about 114.76 km^2 (15.57% of the total area of the Sangihe Islands Regency).

In general, the average monthly air temperature at the 2016 Naha Meteorological Station measurements is 27.8°C, where the lowest air temperature is 20.0°C (in March), and the highest air temperature is 34.0°C (in July). Rainfall in a place is influenced by climatic conditions, geographical conditions, and the rotation or meeting of air currents. Therefore, precipitation varies monthly. The highest rainfall in 2016 occurred in November, namely 465 mm^3 with 24 rainy days, while the lowest rainfall occurred in March, which was 40 mm^3 with 16 rainy days.

The population of the Sangihe Islands Regency in 2016, based on the population projection, was 130,024 people with 34,040 households and a population density of 176.43 people/km^2. North Tabukan Subdistrict is the most populated with namely 15.15% of the total population in Sangihe. Also, the highest population density is in Tahuna District as the capital of the Sangihe Islands Regency, which is 717.39 people per square kilometer.

3.2 Agricultural characteristics

The condition of the land in the border area of the Sangihe Islands Regency is included in the Dry Land Agroecosystem Zone. In fulfilling rice needs, the area must rely on supplies from outside the region, both from the Regency and Province. Generally, cultivation lands in border areas are used for root crops such as sweet potatoes, cassava, and taro (local tubers specific to the location). The specific condition in this border area is the presence of a local tuber/taro plant called the *daluga* tuber. The tuber is a type of taro tuber that belongs to the Xanthosomasp family, grows wild in swampy areas, and is used by the community as a food reserve.

Horticultural crops, especially vegetables such as chili, tomatoes, eggplant, are cultivated at a household scale through home gardens. This is also true for fruits. Mango, pineapple, banana and orange, coconut, nutmeg, and cloves are the most widely cultivated plantation crops by farmers and are spread throughout the border areas. The use of coconut by farmers is only limited to making copra and household needs.

4. Potential development of local tubers

4.1 Varieties of local tubers

Local tubers, as sources of non-rice carbohydrates, are specific food crops for people in border areas, with the potential to be developed as alternative food ingredients to support food security. There are several types of local tubers in the North Sulawesi Province, especially those in the border area of the Sangihe Islands Regency. Types of local tubers are cassava, sweet potato, and taro. These tubers are spread over 15 sub-districts. Production potential recorded in 2016 was for cassava with a harvested area of 302.5 ha, production of 1,210 tons; sweet potato with a harvested area of 186 ha and production of 806 tons and taro with a harvested area of 213.5 ha and production of 759 tons. South Tabukan Sub-district is the largest contributor to production, with around 49.92% [14].

4.1.1 Cassava (Manihotutilissima)

Cassava (*Manihotutilissima*) is a tropical and subtropical plant of the Euphorbiaceae family. It has a taproot and some branch roots that enlarge into root tubers. It is considered an adaptable plant that grows in various tropical agro-climates and does not demand a specific climate in its growth [15]. It was also stated by [16], that "cassava is a wind-resistant plant and will thrive under conditions of low soil fertility". This type of plant can grow in any place, especially in the tropics with full sun throughout the year, and has high adaptability to various soil conditions. Santoso and Radjit [17] stated that cassava production centers are usually located on dry land on alkaline soil and acid soil which are poor in organic matter and macro and micronutrients with weed disturbances. Because the cassava plant has wide adaptability, it can live and produce on land under these conditions. This aspect is due to the nature of the cassava plant which is very efficient at absorbing nutrients in the soil.

According to the Center for Agricultural Information and Information Systems [18], cassava is a substitute for rice with an important role in supporting the food security of a region. It has a fairly complete nutritional content. The chemical and nutritional contents of cassava are carbohydrates, fat, protein, dietary fiber, vitamins (B1, C), minerals (Fe, F, Ca), non-nutritive substances, and water. Besides, cassava contains non-nutritive tannin compounds [19]. Furthermore, [20] stated that cassava has a fairly good nutritional value and is indispensable for maintaining a healthy body, as food, especially as a source of carbohydrates, but poor in protein. The nutritional content of cassava can be seen in **Table 1**.

Cassava, as a source of carbohydrates, can be used as animal feed and industrial raw materials. Therefore, the development of cassava is crucial in efforts to provide non-rice carbohydrate foods, diversify local food consumption, develop product processing and agro-industries as well a source of foreign exchange through exports and efforts to support increased food security and food independence. Although cassava is a source of carbohydrates, the yield of the plant at present is not optimal. Cassava is usually only boiled, fried, or processed into chips. Various variations of food can be produced from cassava. Cassava flour can be used to replace wheat flour.

Component	White Cassava	Yellow Cassava
Energy (Cal)	146	157
Protein (g)	1.20	0.80
Fat (g)	0.30	0.30
Carbohydrates (g)	34.70	37.90
Ca (mg)	33.00	33.00
P (mg)	40.00	40.00
Fe (g)	0	0.70
Vitamin A (SI)	0.70	386
Vitamin B1 (mg)	0.06	0.06
Vitamin C (mg)	30	30
Water (g)	62.50	60
Edible part (g)	75	75
Source: [19].		

Table 1.
Chemical composition of cassava per 100 g.

4.1.2 Sweet potato (Ipomoea batatas L.)

Sweet potato has great potential as an alternative food. It is quite popular in Indonesian society, especially in the eastern region, which uses sweet potato as a staple food. Sweet potatoes are a very healthy and very good food ingredient. This condition is because sweet potatoes have a high nutritional content of complex carbohydrates, thus, leading to a gradual energy release. Among staple foods, white sweet potato contains the highest calcium compared to rice, corn, wheat, and sorghum. The calcium content can reach 51 mg/100 grams for yellow sweet potatoes [21] (**Table 2**).

Apart from being a source of carbohydrates, the potential of sweet potatoes in the context of diversifying staple foods from local resources is very good. The low price of sweet potato and its affordability at all levels of society is a major factor to encourage business diversification of staple foods other than rice. Sweet potato is a local source of carbohydrates that is used for its root tubers. In Indonesia, sweet

Composition	Content/100 grams				
	Rice	Corn	Wheat	Sorghum	Sweet potato
Calories (cal)	360	361	365	332	152
Protein (g)	6.8	8.7	8.9	11.0	1.5
Fat (g)	0.7	4.5	1.3	3.3	0.3
Carbohydrates (g)	78.9	72.4	77.3	73.0	35.7
Calcium (mg)	6.0	9.0	16.0	28.0	29
Iron (mg)	1.0	5.0	1.0	4.0	0.8
Phosphorus (mg)	140	380	106	287	64
Vitamin B1 (mg)	0.12	0.27	0.12	0.38	0.17
Source: [21].					

Table 2.
List of food ingredients per 100 grams.

potatoes are used as raw material for flour, instant rice, bakpia, donuts, chips, noodles, and pearl rice. Sweet potato flour can be processed into various food products similar to foods made from wheat flour, such as candy, ice cream, bread, cakes, and some soft drinks.

The development of sweet potatoes for various processed products is very perspective because, in addition to the multi-use nature of sweet potatoes, the technology for processing agricultural products is quite advanced in Indonesia. With processing technology, sweet potatoes can be processed into various products such as chips, starch, flour, sauce, jam, chips, croquettes, tape, kremes, brem, getuk, pilus, fried sweet potatoes, boiled sweet potatoes, and sweet potatoes. In the form of processed products, sweet potatoes can be upgraded to the equivalent of rice. Sweet potato is also a raw material for the food and non-food industry which is more successful. The success of the food diversification program will reduce dependence on imported rice [22].

4.1.3 Taro (Colocasiaesculenta L)

Taro is a year-round plant. It can grow in various areas, both natural and farmed. This plant, widely grown in rural areas, is usually used as a food substitute for rice, snacks, and even just allowed to grow [23]. In the border area of the Sangihe Islands Regency, there are two types of local taro specific to the location, namely Daluga tubers, and Kole Rea tubers. These two taro tubers are used by some people as a staple food to replace rice.

4.1.3.1 Daluga tubers

Daluga tubers are included in the taro tuber group in the Araceae family. This tuber is a commodity that has important prospects and has high economic value compared to other types of tubers such as sweet potatoes and cassava. Taro is an important food source because the tubers are foodstuffs that have good nutritional values. Daluga tubers can be harvested after about 10 months to 3 years. Bulb weight is quite high, on average 2–5 kg per tuber. Daluga lives well in places that are quite watery such as riverbanks or marshy land and are somewhat protected from the sun. Daluga reproduces by seeds or vegetative [24]. In some border areas, the potential of this tuber is quite promising, but rice is increasingly known to the public. This tuber is no longer cultivated, only planted wildly and not maintained. The highest nutrient content in taro is starch, although it varies between types of taro. Besides being used as a source of carbohydrates, taro tubers can also be used as a functional food because of their high oligosaccharide content [25]. Ref. [26] stated that, when viewed from the nutritional content, taro tubers are considered healthy food commodities and the level of safety lies in their low carbohydrate content (22.25%), reduced sugar (0.87%), and starch content (24, 25%, 11%). The results of the study [9] showed that daluga tuber contains a fairly high carbohydrate with 32.53%, and the flour contains fat of about 23.32% and starch content of 48.86% (**Table 3**).

4.1.3.2 Cole Rea Bulbs

This type of taro for the people on the border of Sangihe is known as kolerea which means looking for sweet potatoes. It has white tubers. The border area of this population is large compared to daluga tubers. In addition to taro kolerea, there is also taro with purple leaf stalks known as bete retraction. The level of community consumption of taro colerea is still high because of the easiness of cultivation and

Parameter	Daluga bulbs	Daluga flour
Water	63.86	1.11
Protein	0.64	1.97
Fat	1.43	23.32
Carbohydrates	32.53	48.86 (starch)

Source: [9].

Table 3.
Nutrient content of tubers and daluga flour.

maintenance. Thus, some community members cultivate this taro intensively in the yard and the garden. This plant is intensively cultivated by paying attention to the nursery and its maintenance.

4.2 Basic food commodities

The results of a study conducted by [27], concerning Location Quotient (LO) analysis, reported that the food crop in the agricultural sector, especially local tubers (cassava and sweet potato) in the border area of Sangihe Islands Regency, had a location quotient (LQ) value >1. Cassava has a value of 9.1, while sweet potato has a value of 12.64 (**Table 4**). With these values, the food crop commodity can meet the needs in the border area of the Sangihe Islands Regency and is expected to encourage the growth of other economic sectors so that it can increase the economic growth rate of the region.

Location Quotient (LO) analysis, at the sub-district level in the border area of the Sangihe Islands Regency, shows the LQ value of >1 for cassava spread over several sub-districts, with LQ values of 1.39 in South Central Tabukan, 1, 06 in South Southeast Tabukan,1.39 in central Tabukan, 1.4 in Manganitu, 1.05 in West Tabukan, 1.18 in North Tabukan and 1.03 in Kendahe. For sweet potato commodities, the LQ values were as follows: South Tabukan District had 1.45, South Central Tabukan with 2.09, South Southeast Tabukan with 1.11, Central Tabukan with 2.09, Manganese with 2.11, Tahuna with 1.04, East Tahuna with 1.72, West Year with 1.58, North Tabukan with 1.77 and Kendahe with 1.73. The value of taro commodity in South Manganitu District was 1.56, 2.29 in Tatoaren, 1.26 in Tamako, 1.09 in South Tabukan, 2.21 in Central Tabukan, 1.14 in Tahuna and 1.31 in Tahuna Timur (**Table 5**). This result shows that the yields of the three food crop commodities (cassava, sweet potato, and taro) make them the basic commodities that can meet the needs in the border areas of the Sangihe Islands Regency.

The production of tubers recorded in 2016 was cassava, with a harvested area of 302.5 ha, and a production of 1,210 tons, followed by sweet potato harvested with

Commodity	Districts Sangihe Islands	Province North Sulawesi	Location Quotient (LQ)
	Production (ton)	Production (ton)	
Cassava	9766.70	279.22	9.1
Sweet potato	9441.87	192.43	12.64
Total	22,456.79	5,785.66	

Source: [27].

Table 4.
Location quotient (LQ) production of food crops (local tubers) level districts.

Number	Districts	Commodity							
		Cassava		Sweet potato		Taro	
		Production (ton)	LQ	Production (ton)	LQ	Production (ton)	LQ
1	Manginitu Selatan	42	0.86	22.5	0.69	48	1.56
2	Tatoaren	12	0.86	—	—	20	2.29
3	Tamako	16	0.93	45	0.59	32	1.26
4	Tabukan Selatan	604	0.97	400.5	1.45	428	1.09
5	Tabukan Selatan Tengah	72	1.39	22.5	2.09	24	0.74
6	Tabukan Selatan Tenggara	52	1.06	36	1.11	24	0.78
7	Tabukan Tengah	44	1.39	4.5	2.09	24	2.21
8	Manganitu	108	1.4	45	2.1	24	0,5
9	Tahuna	28	0.69	36	1.04	29	1.14
10	TahunaTimur	28	0.57	8	1.72	29	1.31
11	Tahuna Barat	52	1.05	49.5	1.58	12	0.39
12	Tabukan Utara	112	1.18	85.5	1.77	20	0.34
13	Nusa Tabukan	—	—	9	0.54	48	3.08
14	Marore	4	1.15	2	0.86	2	0.91
15	Kendahe	36	1.03	40.5	1.73	4	0.18

Source: Result of data analysis (2021).

Table 5.
Location quotient (LQ) production of food crops (local tubers) district level.

an area of 186 ha and a production of 806 tons and taro with an area of 213.5 ha and a production of 759 tons. South TabukanSubdistrict was the largest contributor to production, which was around 49.92% [14]. This condition opens opportunities for its development and it is hoped that the farming system of the three commodities will encourage the growth of other economic sectors to increase the economic growth rate in border areas.

4.3 Value chain analysis of local bulb farming

The results of field observations of various tuber products in the border areas of the Sangihe Islands Regency showed that the economic value is still dominant only from primary products in the form of wet tubers, even though the economic value will be several times higher if there are additional productive activities in each channel such as large-scale product processing, economy, structuring the marketing system, as well as packaging processed tuber products [28] stated that this integration pattern between production and land productivity can be increased or farmers' incomes can also increase and be more resistant to various risks, such as season, price, and income generation. By using the production data of tubers (cassava, sweet potato, and taro) in 2016 in the border area of the Sangihe Islands Regency in 2016, if only half of these primary products were to take a value chain approach with an added value of IDR 4500/kg, there would be an increase in production value of IDR 2.72 billion (cassava), IDR 1.81 billion (sweet potato) and 1.70 billion (taro) with a total value of IDR 6.20 billion. The income of farmers from tubers farming with primary products in the form of wet tubers is only around IDR 2000–IDR 3000/kg with potential productivity of 20 tons/harvest/ha so that a production value of around IDR 40 million–IDR 60 million with a net income of around IDR 20 million–IDR 30 million/ha/year. Through the value chain approach at the tuber farmer level as above, farmers will get an additional production value of approximately IDR 30 million–IDR 50 million/ha/year.

It is an indication that the current condition of the tuber product value is only in the form of wet tubers and it is necessary to immediately switch to other, more profitable products. The current condition of the products produced is still dominant for local needs. Efforts to increase income from tuber farming, the processing of tuber products are relevant options. The development of tuber farming in North Sulawesi Province, especially in border areas, is classified as crucial and has the opportunity for exportation.

Generally, tuber farmers sell their products only individually directly to collectors or consumers. This is an activity to shorten the marketing chain with a collective sales system. The difference in prices in the form of wet primary products from village/sub-district collectors with consumers or manufacturers is usually a price difference of around IDR 1000–IDR 2000/kg. For tuber products that have been processed (flour form), the difference will be even greater. If farmers in one village can produce 200 tons/year with a price difference of IDR 2000/kg, then farmers in the village have lost their income of IDR 400 million/village/year. Therefore, tubers farmers have the opportunity to generate additional collective income of around IDR 300 million–IDR 350 million/village/year. If the farmer has 20 tons of wet tubers, there is an opportunity for additional income per year of IDR 20 million–IDR 25 million/year.

5. Tuber plant technology and support innovation

Technological innovation plays an important role in agricultural development. Innovative technology is produced through research activities, both in the context of improving the existing technology (indigenous technology) and creating completely new technology. Some of the superior varieties of tubers that have been produced by the Agricultural Research and Development Agency are as shown in **Tables 6** and 7.

In Indonesia, tubers are used as raw materials for flour, instant rice, bakpia, donuts, chips, noodles, and pearl rice. Flour derived from tubers can be processed into a variety of food products similar to food ingredients made from wheat flour, such as sweets, ice cream, bread, cakes, and some soft drinks. Currently, the use of wheat flour as a substitute for wheat flour is not a new development. The development of root crops for various processed products is prospective, because of its multi-purpose nature. Taro tubers can be processed into various products with nutritional value. Products that can be produced from taro tubers can be grouped into categories that include the development of (1) products from fresh tubers, (2) intermediate products, (3) ready-to-cook products, and (4) ready-to-eat products of fresh tubers such as taro flour, taro chips, and traditional food products [32]. Flour processing is the best choice because: (1) flour is a product that is practical to use, so that it can be processed directly into instant food or as raw materials of other food products, (2) flour-processing technology is very easy to adopt and apply at low cost, so that small to medium-sized businesses can develop this business (3) flour easily fortified with the necessary nutrients such as vitamins and minerals and, (4) people have become accustomed to consuming food derived from flour.

Varieties	Productivity (ton/ha)	Harvest age (month)	Pest/disease resistance
Adira-1	22	7–10	Somewhat resistant to red mites; resistant to leaf blight, resistant to wilting
Adira-2	22	8–12	Fairly resistant to red mites; wither
Adira-4	35	10	Enough red mites
Malang-1	24.3–48.7	9–10	Fairly resistant to red mites, tolerant of leaf spot, wide adaptability
Malang-2	20–24	80–10	Slightly sensitive to red mites, tolerant of leaf spot and leaf blight
Darul Hidayah	10–21	8–12	Slightly sensitive to red mites and fungal rot
UJ-3	20–35	8–10	Resistant to leaf blight bacteria
UJ-5	25–38	9–10	Resistant to leaf blight bacteria
Malang-4	39.7	9	Somewhat resistant to red mites, adaptive to sub-optimal nutrients
Malang-6	36.41	9	Somewhat resistant to red mites, adaptive to sub-optimal nutrients

Sources: [29–31].

Table 6.
Description of several superior varieties of cassava by the research agency and agricultural development.

Varieties	Release year	Productivity (ton/ha)	Harvest age (month)	Characteristics
Muara Takus	1995	30–35	4.0–4.5	Resistant to scab/scab disease, good tuber shape, high tuber dry matter weight, suitable for planting in dry land and paddy fields
Cangkuang	1998	30–31	4.0–4.5	Somewhat resistant to lanas pests, resistant to scurvy, good shape of tubers, high dry matter weight of tubers, a high percentage of tuber weight, suitable for planting on dry land or rice fields after rice, which is not very fertile
Sewu	1998	28.5–30.0	4.0–4.5	Slightly resistant to lanas pests, resistant to scabies, good tuber shape, medium-dry matter weight, suitable for planting on dry land or rice fields after rice
Boko	2001	25–30	4–4.5	Moderately resistant to boleng/lanas/borers and resistant to leaf rollers, tolerant of scabies and leaf spot.
Sukuh	2001	25–30	4–4.5	Somewhat resistant to boleng/lanas/borers and leaf curlers, resistant to scabies and leaf spot
Jago	2001	25–30	4–4.5	Somewhat resistant to boleng/lanas/borers and leaf rollers, moderately resistant to scabies and leaf spot.
Kidal	2001	25–30	4–4.5	Slightly durable with holes/lanas/borers and leaf rollers, resistant to scurvy and leaf spot bercak

Sources: [29–31].

Table 7.
Description of several superior sweet potato varieties of agricultural research and development agency.

Derivative products of taro flour can be used as dodol, various wet and dry cakes, noodles, cheese sticks, bread, breakfast meal, analog rice, cookies or biscuits, and sauces [33]. From a fresh state, tubers can be processed into a variety of ready meals or snacks, dried sawut or gaplek, chips, starch, and tuber flour. Many ready meals are made from fresh tubers, such as pilus, cakes, croquettes, enyek-enyek, getuk, or various kinds of cakes [34].

Fresh Taro daluga can be processed into a variety of products including chips, dodol, brownies, dried mustard, noodles, and various other wet cakes. Dodol is one type of processed food that is classified as semi-wet food because it has a water content of 10–40% with a water activity of 0.65–0.90 it has an elastic and dense texture [33]. The product is easy to process and can increase added value and diversify the product. Tubers talas in the form of flour have better nutritional composition than rice. Taro flour contains higher protein and lower fat than rice. The fiber content of taro is also quite high and very good for maintaining the health of the digestive tract. Taro flour is classified as smooth and easy to digest. It is useful for the manufacture of pastries, cakes, bread, and noodles [35]. Processing of taro flour products is expected to minimize losses due to fresh taro tubers not being sold out when over-harvest production. Besides, taro flour can be used as a substitute for processed food products such as sweet bread [36]. The use of taro can increase the economic value in the form of flour and taro starch as well as the shelf life of taro production. Taro starch can be used as a new type of starch and an alternative companion or substitute for wheat. Processing taro tubers with taro flour raw materials is still limited because taro flour is not available on the market [37]. One stage of the flour-making process is drying, where the drying temperature affects swelling *power,* solubility, and myelography properties.

6. Local tuber development challenges and problems

Local food in the border area, especially the archipelago area, is different and has its characteristics compared to local food in non-island areas [38]. The challenges faced in the development of local tubers in the border area are based on potential analysis with the approach of border areas and value chains. Land use, the potential of existing land has not been utilized optimally. The land is generally dominated by dry land and some swamps, so there are constraints on development and utilization in trying to farm. Land capability implies land carrying capacity. Land capability is the quality of land that is assessed with the understanding of a compound identifier of land and the value of land capabilities is different for different uses. Concerning the fulfillment of human needs, the ability of land is described in the understanding of land carrying capacity [39].

The climate and weather conditions in the border region are erratic and often capricious. During the northern wind season, wind speeds can reach 40 mph with seas surges. These natural conditions result in residents or communities on the border experiencing shortages of foodstuffs. There are generally border areas included in the criteria of poor villages, with growth tending to be slower compared to the surrounding villages [40]. Some factors that cause the slow growth of villages in the border areas include (a) no thorough identification regarding the socio-economic potential of the people in the border area, essentially a supporting factor for the resilience of the people in the border area; (b) the weak ability of social and economic services of the people in the border area compared to the number of people to be served; and (c) the lack of evenly distributed social and economic services in the border areas seen based on location or spatial distribution; (d) lack of community motivation in improving the household economy through crop cultivation efforts.

Based on the Sangihe Islands, Human Development Index in 2014 was 66.82, lower than the average HDI of North Sulawesi that reached 69.96. This condition is partly seen from the low level of education of the population aged 15 years and above, who are only elementary school graduates (52%) [13]. People in the border region have a perception of the prospect of developing root crops (cassava, sweet potatoes, and taro), although the level of preference for rice is higher. The people of the city see it as the foodstuff of the weak economic class or rural communities. On the contrary, rural communities see it as a commodity of high social value, as it is usually served in traditional parties, such as weddings, chief appointments, welcoming guests, and death.

Generally, the management system of tuber farming in the border area of the Sangihe Islands regency is dominantly conventional. The factors that influence farmers' decision to adopt technology are the direct benefit of technology in the form of relative benefits, conformity of technology to socio-cultural values, and ways and habits of farming [41]. The economic value of tuber farming products will be higher if every sub-part of the agribusiness system can carry out productive activities to create benefits and employment opportunities. To increase the income of tuber farming, the processing of tubers products becomes a relevant option. The root products also have the potential to become feed for livestock development. So far, the mainstay of the economic value of the tubers is still very dependent on the primary product [26].

Given the ownership of assets, the farming community is relatively small, individual actions in business development will be very difficult in reaching optimal benefits. Therefore, the development of business in the future that is of maximum added value needs to immediately take collective action in the sale of proceeds, purchase of production facilities, investment funds, and access to new technology information and business partners.

Some of the main problems that must be addressed in the development of local food value chains in the border areas include:

1. Inconsistent regulatory/policy support to improve commodity competitiveness.

2. Potential food insecurity and malnutrition for people in isolated areas.

3. Low productivity of local quality food.

4. Unavailable downstream industry players.

5. The potential of the local market has not been optimal so market access is still limited.

6. Still the low quality of human resources, weak institutional both at the level of the main actors and business support institutions in the value chain of tuber development.

7. Limited availability of field extension workers.

8. Weak coordination and partnership between government-private actors.

9. Availability of infrastructure that is not optimal.

The main factor that weakens agricultural businesses, including the development of tuber commodities, is that farmers' economic institutions do not have strong intentions to build [42]. Stated that the lack of functioning as agricultural institutions were partial since the establishment of these institutions was not carried out in a participatory manner, where farmers as beneficiaries and placed as actors running these institutions.

7. Strategic steps in local tubers development

Facing an era of globalization and free competition, small agricultural-based industries need attention to increasing the added value of local food products as the economic center of communities in the border region. Strategic steps that can be taken in the development of local tubers to increase production and productivity in the border areas of Sangihe Regency include:

a. Improvement of regulations/policies that support the business climate and infrastructure:

- Central government support through accelerated development program in the outer border areas of the island.

- Banking support on credit base rate for businesses.

- The policy of the Ministry of Agriculture to facilitate the certification of geographical indications as a form of protection of the authenticity of agricultural products of an area can have the opportunity to improve the competitiveness and marketing of food products and change public consumption patterns.

- Allocating budgets for tasks and functions in the agricultural sector in the border region.

- Increasing the motivation of farmers in cultivating local tubers.

- Increased community preference for local tubers-based food.

- Providing its main infrastructure access from all industrial centers to the city or market including improved transportation services.

- Limiting the transfer of agricultural land to settlements or roads.

b. Institutional strengthening of organizations and supporting the development of local tubers:

- Improvement of support agencies involved in the development of local tubers.

- Improving the ability of farmers in carrying out cultivation technology and post-harvest handling and processing of yields.

- Changing the mindset of farmers who are still oriented to meet the needs of their own families and have not been oriented to commercial businesses.

Local Tubers Production and Value Chain: Evidence from Sangihe Island Regency, Indonesia
DOI: http://dx.doi.org/10.5772/intechopen.103711

- Increasing the number of agricultural extension workers so that farmers get information about the latest technology.

- Providing processed industries and product packaging.

- Grow organizations that can represent farmers or groups of actors in the value chain.

- Involves the role of indigenous institutions in encouraging the cultivation of local tubers.

c. Development of patterns of cooperation and partnership between government-private and community.

- The development of local tubers cannot be done individually, it must be done in an integrated manner, requiring the participation of businesses that understand the production process and market information.

- Increasing the role of local governments in supporting problem-solving in farmers, collectors, traders, and processed industries.

- Improving the role of society including increased knowledge/awareness and increased income.

- Improved partnership. The implementation, synchronization, and cooperation between all stakeholders in the development of food consumption including the development of food processing technology.

- Optimizing the system of coordination and partnership between supporting institutions due to the ego of sectoral interests.

d. Research, development, and innovation regarding cultivation technology and development of derivatives. Local tubers in the border area of Sangihe Islands regency have not been considered important commodities, while in some areas in Indonesia, they are used as food and non-food raw materials, such as noodles, fried cassava, dessert, confectionery, soy sauce, flour, wine, vinegar, nata de coco, and others. Even lately with a limited supply of energy sources, sweet potatoes are explored to be an updated alternative energy source, including converting sweet potatoes into bioethanol. Meanwhile, in the border area, exploration of the utilization of local tubers is still very far behind. The current condition of the majority of local tuber utilization is still limited to the main food sources only, so efforts to diversify local tuber derivative products have not developed optimally.

8. Conclusions

The border area is not only understood as a geographical concept of the region that is directly adjacent to other countries but also a strategic area that nationally concerns the lives of many people, whether or not it is reviewed for political, economic, social, cultural, and environmental and security defense interests. Local tubers in the border area of the Sangihe Islands Regency have the potential as base commodity plants and support technological innovations available to be developed both in terms of cultivation and industrial products with high economic value.

Some problems faced in the development of local tuber crops in the border area are, the potential for untapped land, climate and weather conditions in the border region that are erratic and often capricious, less motivation of farmers in improving the household economy through the business of cultivating crops and ownership of assets of farmers which is relatively small; then individual actions in business development will be very difficult in reaching optimal added value. Strategies that can be done in the development of local tubers to increase production and productivity in the border area of Sangihe Regency, among others, are regulations/policies repairing that support the business climate and infrastructure, institutional organization, development patterns of cooperation and partnership between government-private and community and research, development and innovation on cultivation technology and development of processed products with economic value.

9. Development policy implication

The government's development efforts and strategies include accelerating the economic growth of border areas through people's economic base with the availability of adequate infrastructure, conducive and constructive political stability to support economic growth in the region. This condition can be achieved through community empowerment by increasing the role and participation of communities in border areas and improving development management performance through improvement of the quality of government officials so that they can become facilitators of border area development.

For this reason, it is recommended that government officials as development policymakers should be able to encourage the management of natural resources in border areas based on superior commodities in increasing production and value chains. The development of local root crops is the main recommendation to improve food security which is still low in addition to improving the welfare of people with low purchasing power. To increase the productivity of local tuber farming, it is necessary to introduce superior seeds on time, including the provision of agricultural production facilities supported by the application of cultivation and post-harvest technology. Meanwhile, to improve the value chain, namely to strengthen the existence of farmer groups so that it not only increases bargaining power but also reduces transaction costs in marketing. Meanwhile, improving vertical coordination is carried out by establishing a network of partnerships with market players and fulfilling contractual agreements in profitable markets.

Local Tubers Production and Value Chain: Evidence from Sangihe Island Regency, Indonesia
DOI: http://dx.doi.org/10.5772/intechopen.103711

Author details

Agustinus N. Kairupan*, Gabriel H. Joseph, Jantje G. Kindangen,
Paulus C. Paat, August Polakitan, Derek Polakitan, Ibrahim Erik Malia
and Ronald Hutapea
Assessment Institute for Agricultural Technology of North Sulawesi, Manado,
Indonesia

*Address all correspondence to: audikairupan@gmail.com

IntechOpen

References

[1] Atem NN. Persoalan Kerawanan Pangan pada Masyarakat Miskin di Wilayah Perbatasan Entikong (Indonesia-Malaysia) Kalimantan Barat. Jurnal Surya Masyarakat. 2020;**2**(2):94-104

[2] Hermanto. Ketahanan Pangan Indonesia Di Kawasan Asean. Forum Penelitian Agro Ekonomi. 2015;**33**(1):19-31

[3] Winarno B. Dinamika Isu-Isu Global Kontemporer. Yogyakarta: CAPS; 2014

[4] Syarief R, Sumardjo AF. Assessment of food security empowerment modeling inter-state border. Jurnal Ilmu Pertanian Indonesia. 2014;**1**(9):9-13

[5] Priyanto D, Diwyanto K. Agricultural Development in the Borderline Areas of East Nusa Tenggara and Democratic Republic of Timor Leste. Pengembangan Inovasi Pertanian. 2014;**7**(4):207-220

[6] Suhardi. Development of local food-based agro-industry to enhance food sovereignty. In: Proceedings of the National Seminar on the Development of Local Food Source-Based Products to Support Food Sovereignty. Yogyakarta: Agricultural Products Technology study program of Mercu Buana University; 2008

[7] Dewi YA, Hendradi A, Eko Ananto E. Building Food Independence Based on Local Food Utilization in Small Islands and Border Areas; 2013

[8] Sumedi and Djauhari. Strengthening Food Self-Sufficiency: Efforts to Strengthen Food Independence of Small Islands and Border Areas. IAARD Press Agricultural Research and Development Agency; 2015

[9] Lintang M, Layuk, L., Taulu, J., Sondakh, J., Wenas Study Identification Varietas and Utilization of Specific Wild Tubers Location in North Sulawesi as an Alternative Food Source in order to Support Food Security. Final Report on Incentive Research on Improving the Ability of Researchers and Engineers. Ministry of Agriculture; 2012

[10] Priyanto D, Diwyanto K. Pengembangan pertanian wilayah perbatasan Nusa Tenggara Timur dan Republik Demokrasi Timor Leste. Pengembangan Inovasi Pertanian. 2018;**7**(4):207-220

[11] Mulyo JH, Irham J, Widodo AH, Wirakusuma G, Perwitasari H. Food security of farm households in Indonesia's border area, Sebatik Island. International Journal of Engineering &Technology. 2018;**7**(3.30):314-319

[12] Pongtuluran Y. Developing economy in the border of East Kalimantan. Journa SAVAP International. 2013;**4**(4):544-549

[13] Central Bureau of Statistics. BPS North Sulawesi Province. 2015

[14] Central Bureau of Statistics. BPS North Sulawesi Province. 2017

[15] Salim E. Processing Cassava into Mocaf Flour.Yogyakarta: Andi Offset; 2011

[16] Fasinmirin JT, Reichert JM. Conservation tillage for cassava (*Manihotesculentacrantz*) production in the tropics. Soil& Tillage Research. 2011;**113**:1-10

[17] Santoso B, Radjit NP. Optimization of cassava yield using adaptive technology. Food crops science and technology. Food crops research and development bulletin; research institute for various nuts and tubers. Indonesian Center for Food Crops Research and Development, Indonesian Agency for Agriculture Research and Development. 2011;**6**(2). ISSN 1907-4263

[18] Agricultural Data and Information Center, Ministry of Agriculture. Outlook Cassava. 2016. ISSN: 1907-1507

[19] Soenarso S. Maintaining Physical Health Through Food. Bandung: ITB; 2004

[20] Widyastuti E. Characteristics of Bulbs–Tubers. Malang: Brawijaya University; 2012

[21] Direktorat Gizi Depkes RI. List of Indonesian Food Nutrient Composition. Jakarta: Department of Health RI; 2010

[22] Swastika DKS, Hadi PU, Ilham N. Projection of Supply and Demand for Food Crops Commodities: 2000-2020. Jakarta: Agriculture Department; 2000

[23] Sriyono. Making Taro Bulb Chips (*Colocasiagiganteum*) with Variable Frying Time Using a Vacuum Fryer. Laporan Tugas Akhir. Semarang: Diponegoro University; 2012

[24] Layuk PA. Latif dan Adnan. Characteristics and Physicochemical Properties of Daluga Bulb Flour (Xanthosoma sp.). Proceedings of the National Seminar in Jayapura; 7-8 October 2010

[25] Hartati NS, dan T, K Prana. Analysis of starch and crude fiber content of flour several cultivars of Taro (*Colocasia vulgaris* L.). Nature Indonesia. 2003;**6**(1):29-33

[26] Suminarti NE. Nutrient composition of various types of bulbs from tubers and milled rice [dissertation Faculty of Agriculture]. Brawijaya University; 2009

[27] Kairupan AN, Manoppo C. Analysis of economic potential areas based on farming agricultural sector in The Border Area of Sangihe Island Regency. E3S Web of Conferences. 2021;**232**

[28] Kasryno F, Soeparno H. Dry Land Agriculture as a Solution to Realize Future Food Self-reliance. Prospect of Dryland Agriculture in Supporting Food Security. Indonesian Agency for Agriculture Research and Development. Ministry of Agriculture; 2012

[29] Suhartina. Description of Superior Varieties of Nuts and Tubers. Malang: Research Institute For Various Nuts and Tubers; 2005. p. 154

[30] Research Institute For Various Nuts and Tubers. Production Technology of Nuts and Tubers; 2005. Malang: Research Institute For Various Nuts and Tubers, Indonesian Agency for Agriculture Research and Development, Ministry of Agriculture; 2012. p. 36

[31] Wargiono J, Hasanuddin A, dan Suyamto. Cassava Production Technology Supports Bioethanol Industry. Bogor: Indonesian Center for Food Crops Research and Development; 2006. p. 42

[32] Wisnu B. Utilization of Local Food for Diversification of Food Consumption. Processing Technology for Diversification of Food Consumption. Center for Agricultural Post-Harvest and Development, Indonesian Agency for Agriculture Research and Development; 2008

[33] Koswara S. Tubers Processing Technology. Part 1: Processing of Taro Bulbs. 2013. Available from: http://seafast.ipb.ac.id [Accessed: February 25, 2014]

[34] Indonesian Agency for Agriculture Research and Development. Various Processed Bulbs/Agricultural Research and Development Agency. Ministry of Agriculture-Jakarta: IAARD Press; 2012

[35] Gumilang R, Susilo. Test the characteristics of instant noodles made of wheat flour with Taro Flour substitution (*Colocasia esculenta* (L.) Schott). Jurnal Bioproses Komoditas Tropis. 2015;**3**(2):53-56

[36] Pratiwi A, Ansharullah ARB. Effect of Taro Flour substitution (*Colocasia Esculenta* L. Schoott) on sensory value and nutritional value of sweet bread. Journal of Sainsdan Teknologi Pangan. 2017;**2**(4):749-758

[37] Hadi D, Fatarina E, Krisna Anugerah T. Effect of Taro and NaCl ratio with Caco 3 concentration on amylopectin content of taro starch. In: Prosiding SENATEK 2015 Fakultas Teknik, Universitas Muhammadiyah Purwokerto Purwokerto, 28 November 2015. 2015. ISBN: 978-602-4355-0-2

[38] Lintang M, Layuk P, Sondakh SJ. E3S Web of Conferences 232, 02019.*IConARD 2020.* Identification Local Food and Levels of Community Preference at the Border Region of Sangihe Islands District

[39] Notohadiprawiro. Land, Land Use and Spatial Planning in Environmental Impact Analysis. Yogyakarta: Universitas Gadjah Mada Press; 1987

[40] Miniarti L. The potential role of social and economic infrastructure and facilities in the development of border areas in Gunung Kidul Regency, D.I. Province. Yogyakarta. Jurnal SMARTek. 2010;**8**(1):72-82

[41] Indraningsih KS. The effect of extension on farmers' decisions in adopting integrated farming technology innovations. Jurnal Agro Ekonomi. 2011;**29**(1):1-24

[42] Budi GS, Aminah M. Dominant factors in the formation of social institutions. Agro-Economic Research Forum. 2009;**27**(1):29

Section 2

Sustainable Development

Chapter 5

Brazilian Coffee Sustainability, Production, and Certification

Laleska Rossi Moda, Eduardo Eugênio Spers,
Luciana Florêncio de Almeida
and Sandra Mara de Alencar Schiavi

Abstract

Brazil is the largest coffee producer in the world, being responsible for 40% of world total production, 69.9 million bags in 2021. Due its major production and exportation role in the global coffee market, Brazil has been also recognized for its commitment with quality and social-sustainability parameters based on voluntary sustainability standards (VSS) and geographic identification (GI). Despite higher prices at the final market and some changes toward more sustainable production models, certification is not a panacea for sustainability. In that sense, the governance of certification and standards along the value chains plays a central role. Brazil, as the largest coffee producer and exporter, has also a great potential regarding coffee GI, which can lead to differentiation strategies and economic benefits for small farmers, contributing also to sustainable production and cultural and environmental protection. However, the existence of economic and social barriers plays salient challenges for farmers to meet the quality standards as well as GI protocols among other market compliance tools, in addition to the correct value appropriation arising for quality sustainability adopted strategies by coffee farmers in Brazil.

Keywords: coffee, Brazil, sustainability value chain, voluntary sustainability standards (VSS), geographic identification (GI)

1. Introduction

Sustainability has been a crescent worldwide topic in this past decades, especially after the 2000s, as climate change, biodiversity loss, soil erosion, water crisis, among other challenges, are becoming more evident, leading to a crescent concern in the minds of consumers [1, 2]. In food systems (e.g., coffee value chain), sustainability is especially visible: in order to attend the growing population and to avoid economic and social impacts, food production should increase by 70% until 2050 [3, 4]. Hence, the big challenge is how to produce more without destabilizing the ecosystems on which we depend [5].

For coffee, the chain is faced with many challenges, such as water pollution, soil erosion, biodiversity loss, among other climate-related problems, and social impact. Coffee is one of the most traded commodities of the world, but the production is mainly done by millions of small farmers around the world who depend on coffee for their livelihood. Thus, climate change may affect the production areas of coffee and the livelihoods of the producers [4, 6, 7], increasing the concerns for sustainability in the chain.

According to Baumgärtner & Quaas [1], sustainability, per se, can be understood as a normative notion as to how humans should act toward nature and how they are responsible in relation to the people around them and the future generations. The concern with preserving the natural resources for the future is not just a concept of the modern human, but it was present since the Neolithic Revolution and later in many populations around the world. The topic was also studied by economists for a long time, since the shortage of resources is of central concern to the science [8, 9].

Yet the term "sustainability" became popular in policy-oriented research, with the concept of sustainable development, a common goal for society in the twenty-first century, introduced in 1987 by the report Our Common Future, also known as the Brundtland Report [10]. The ideas presented were also later discussed in the United Nations Conference on Environment and Development in 1992—known as the Rio Summit—where a consensus and commitment of the academia were agreed in engaging in development and environmental problems [11].

Based on these ideas, sustainable development can be defined as "the development that meets the needs of the present without compromising the needs of future generations to meet their own needs," and highlighted that, while environmental concerns are important, welfare and intergenerational equity should also be discussed. Thus, sustainability is not just about the environment, but has two more dimensions—economic and social. It is a multidisciplinary subject, and it is the intersection of these three dimensions that also allows the inclusion of socioeconomic factors, besides the environment aspect [9].

The social sphere is about improving poverty and having social inclusion; the economic sustainability regards perduring of renewable and nonrenewable resources of production system in the long run and the economic growth; lastly, environmental aspects are related to protection and conservation of living being (e.g., humans, animals, and plants) existing on Earth [12–15]. This three-dimensional quality of sustainability is also embodied in the definition of the concept by the United Nations in its Sustainable Development Goals, recognizing that social improvement should walk alongside economic growth, while "tackling climate change and working to preserve our oceans and forests" [16].

According to Hajian & Jangchi Kashani [13], sustainability can also be seen in a weak or strong meaning. The former is based on an economic value, the resources are goods with capital value, while the latter sees resources as natural goods and services it delivers, based on biophysical principles, considering some functions that the environment does for humans.

Despite the definitions above, it is good to note that the term "sustainability" does not have an extremely clear meaning. According to Pretty [17], since the Brundtland Report, there have been over 70 definitions of sustainability, each in a subtle way that enhances different goals, values, and priorities. For example, there can be different types of visions of sustainability depending on from whose eyes we are looking through (e.g., people in underdeveloped country and developed countries) and the time period of the action, such as how many years are we talking about the future in terms of generations [13]. So, even with the three-dimensional diagram (economic, social, and environment) of sustainability already consolidated, we can still have some variations in the actions and challenges faced in different areas.

In the agriculture, for instance, one of the most important challenges regarding sustainability is how to reach food security in the future—that is, how to feed the growing population of the world—while facing climate changes, as appointed in the beginning of this chapter [18, 19]. Besides this goal, sustainability is also generally associated with economic viability for farmers, environmental conservation, and social responsibility. The goal is how to maintain or increase the production of goods, thinking about the economic viability fir farmers and food security, while working on the conservation of the resources, such as water, soil, and biodiversity [20].

To reach that, the sector already invested in some standards alongside the chain of certain agricultural product, in an attempt to cover the whole value chain from farmer to consumer [20]. According to Bager & Lambin [6], for the coffee sector, companies normally rely on the adoption of combined codes of conduct, voluntary sustainability standards (VSSs), corporate social responsibility (CSR) programs, direct relations with producers, and so on, to address the challenges of sustainability. It is good to note that the sector is also one of the models regarding sustainable actions, with third-party certification standards being widely used as VSSs, although internal standards and various supply chain interventions are gaining attention on the last years [6, 21, 22]. Other forms of addressing sustainability also include direct trade, single origin, and value chain transparency [6].

Brazil is the largest coffee producer in the world, being responsible for 40% of world total production, 69.9 million bags in 2021. The country is also the largest green coffee exporter, with 45.7 million bags, or 32% of total exports [23]. As for differentiated coffees, which includes sustainable certified ones, the country exported 7.7 million bags in 2021 [24], mainly to the United States, Germany, Belgium, Italy, Japan, and United Kingdom. That amount represents a 50% increase in comparison to 2017.

The enrichment of Brazilian coffee in quality and sustainability parameters over the years has positioned the country as an international reference for institutional and private strategies toward agricultural best practices aligned with sustainability goals.

This chapter aims to exploit the quality-sustainability-led strategies largely adopted by multiple stakeholders at the Brazilian coffee chain as a response for local and global demand for guaranteeing high quality for consumers along with fair prices and quality conditions for coffee famers.

Due to the history and importance of sustainability in the coffee sector, this chapter aimed to give an overview on how this theme has been worked on the coffee value chain in recent years and the possible lessons we can get of that. To reach that, the chapter was divided in the methodology, followed by the findings that contemplated broad aspects of coffee production and demand, as well as the specific aspects of sustainability in these topics. It also included the topic of the standards, certification, and governance regarding sustainability in coffee. Lastly, the authors presented the key findings of this study.

2. Methodology

The method used was a qualitative review of the academic literature and private reports on the coffee value chain and sustainability, based on the importance of the publications. It applied a set of key search terms in two scholarly electronic databases (Web of Science and Science Direct) and on Google Scholar in January–February 2022 to identify relevant papers. The string of key search words used were combinations of "coffee" and "sustainability," "production," "demand," "green," "certification," "standards," "voluntary sustainability standards," 'Designation of Origin," and "Geographic Indication." It was searched within the abstract, title, and keyword database categories of original research papers published in peer-reviewed English and Portuguese language academic. These articles were then selected based on the relevance in the platform's journals. It was also included relevant reports in the coffee sector by instructions and actors such as the International Coffee Organization (ICO), the Global Coffee Platform, and The Economist. Finally, statistics and figures about the sector were obtained from sectoral reports and official databases, such as the Production, Supply & Distribution Online Database, from the United States Department of Agriculture (USDA) and the Brazilian Coffee Exporters Council (CECAFÉ).

3. International coffee demand

Coffee is a multibillionaire business and one of the most traded commodities of the word, involving thousands of companies and millions of coffee growers [6, 25]. According to ICO [26], since the 1990s, coffee production has had an increase of 60%, while the value of exports has more than quadrupled from USD 8.4 billion in 1991 to USD 35.6 billion in 2018, thanks to the rise in consumption and value in the chain. Regarding the production, it is condensed in more than 60 countries in the coffee belt (between the Tropics of Cancer and Capricorn), with Brazil begin the largest producer (33–35%). Yet, most of these countries remain marginal actors, with the international trade of processed coffee dominated by a small number of actors that capture a large value share of the global value chain (GVC), such as members of EU and North America. This is also reflected in the consumption, with the top consumers being mainly developed countries, such as the United States, Germany, Japan, Italy, and France (with the exception of Brazil, the second largest consumer) [26, 27].

This led to some implications in the GVC: today, the coffee value chain is characterized as a buyer-driven chain, where roasters and multinational companies hold the power to coordinate and impose control on the actor in the chain. In this case, while these buyers are subjected to sophisticated institutional regulation within their home countries, they can still exercise power on the producer's end, which can affect their livelihoods and environment. This has led to concerns among consumers and NGOs, who hold large companies accountable for their impact on the environment and laborers. This was especially true in the last years, due to high fluctuation on prices and increase of production costs, caused by climate changes and, since 2020, global chain disruption by the COVID-19 pandemic [26, 28, 29].

This increasing concern for sustainability by consumer (especially in tor consuming countries) is a trend occurring in all of GVCs and has led governments and companies to take action in addressing this matter and meet stakeholders expectations—also increasing income, protecting brand and reputation or differentiation—through the creation standards and regulations [6, 26, 30]. In 2021, the report "An Eco-Awakening" by The Economist Intelligence Unit (EIU) [31] showed an increase of 71% on searches for sustainable goods over the past 5 years (2016–2020) around the world, a trend that continued even during the COVID-19 pandemic. Consumers, waked by the social and environmental concerns, demand each year more actions by companies.

As for the coffee value chain, it is known as a pioneer in the adoption of VSSs, in particular "private" and multistakeholder initiatives, such as the third-party certifications (e.g., 4C, Rainforest Alliance, UTZ, Fairtrade, Organic, etc.) and standards by the private sector (e.g., Starbucks' C.A.F.E. Practices and Nespresso's AAA Guidelines) [6, 29, 32–34].

This trend has been present predominantly since the 2000s. Reinecke et al. [35] showed that the growth rate of coffee sustainable certification was 20% annually. Dietz et al. [36] saw that, while in 2008, the adoption of VSSs was made by 7% of exporters, in 2018, this number increased to 23%, while Panhuysen and Pierrot [37] showed that in the coffee year of 2019/20, 55% of total volume produced was certified with some VSSs. According to ICO [26], investments on sustainability in the coffee chain are estimated to reach USD 350 million annually, showing the great concern of the sector in attending sustainable goals.

As for the Global Coffee Platform [38], an inclusive and important multistakeholder membership organization that seeks sustainability in the coffee sector, the purchase of sustainable coffee (following third-party and second-party schemes) in 2020 reached 16,3 million bags of 60 kg, or 48% of total purchased, for the members. It is good to note that these players include the biggest coffee companies in the world, such as Nestlé, JDE, Melita and Strauss Coffee, which represented a share of 26.6% of world coffee exports and 20.5% of world coffee consumption in 2019/20. The increase in the sustainable coffee purchase between 2019 and 2020 was of 53.1%, with the major origins reported with sustainable coffee purchases being Vietnam, Brazil, Colombia, Honduras, and Mexico. As for the mainly sustainable schemes, 4c certified coffee was the most common (58%), with a high percentage for two or more sustainability schemes, especially triple certification with 4C-Rainforest Alliance-UTZ (10% of the sustainable purchase in 2020).

Most of the biggest coffee companies around the world also have goals to elevate sustainable coffee purchase in the next years, or for a target of "100% responsible coffee in the next decade," such as JDE, Nestlé, and Melita [38]. Despise these optimistic numbers, it is important to point out that not all the sustainable coffee is sold as so: Panhuysen and Pierrot [37] point out that, in 2019, 75% of coffee with VSS schemes were sold as conventional coffee, which might be a challenge for the sustainable coffee sector, affecting price premiums and the differentiation strategies by producers.

4. Voluntary private standards, sustainable certification schemes, and coffee value chain governance in Brazil

Voluntary private standards (VSS) and sustainability certification schemes have taken a central role in discussions about the future of agricultural production and agri-food chains. VSSs are considered important mechanisms to promote sustainability and upgrading in agri-food value chains [39]. In coffee chain, certification schemes are major issues due its importance and impacts in the sustainability as well for farmer's higher incomes [40].

Sustainability coffee certificates in the global coffee industry are present since the world coffee deregulation aiming to guarantee enhanced quality and sustainability in the production regions. The major certifications in the global coffee scenario are Fairtrade (FT), Organic, Rainforest Alliance / UTZ, and the 4C Common Code/Global Coffee Platform (4C/ GCP) [36, 41, 42]. **Table 1** summarizes the main scope and objectives of those VSSs.

As observed in **Table 1**, most common VSSs in coffee value chain comprise the three-dimensional aspect of sustainability—economic, social, and environmental, although in different ways and considering different indicators and measures. Some of them are more focused on one of the dimensions, such as Fairtrade for social aspects, and organic for the environmental dimension. Another important aspect is the scope of VSS in value chain: some of them are more related to one specific segment of the chain (such as organic in production), while others depend more on actions in/from different parts of the chain, such as Fairtrade.

Discussion on the role of VSS for coffee sustainability abound in literature [26, 36, 42–47]. Based on prior studies, Elliott [44] summarizes different impacts of VSS on prices, quality and productivity, income and livelihoods, working conditions, environment impacts, and other aspects, such as markets, training,

VSS	Scope and objectives
FairTrade (FT)	It comprises economic, social, and environmental sustainability for producers, with focus on social aspects, and the strength of labor rights and working conditions. It sets minimum prices and social premia for producers and producers' organizations.
4C Common Code/ Global Coffee Platform (4C/GCP)	It comprises 27 principles across economic, social, and environmental dimensions, aiming to exclude worst practices and increasing sustainability in coffee production and processing.
Organic	It promotes organic farming practices, intended to avoid harmful practices to the environment, prohibiting the adoption of agrochemicals and promoting environmental practices, such as deforestation restriction and soil erosion control.
Rainforest Alliance (RA)/UTZ	UTZ merged with RainForest Alliance in 2018. It establishes standards for responsible production and delivery, aiming to ensure sustainable practices and the integration of biodiversity conservation, community development, labor issues, and agricultural practices.

Source: Based on Dietz [36]; Piao et al. [42].

Table 1.
Major VSSs for coffee and scope.

and capacity building. The findings presented mixed results, which lead to a controversial discussion on the VSS adoption for coffee farmers' income. The same perspective is pointed out for Piao [42] when analyzing the adoption of the 4C system by coffee farmers in Brazil in the perspective of value chain upgrading. The authors had identified five types of upgrading (product, process, functional, social, and environmental) although most of the improvements can be characterized as environmental. Yet, the main gains are associated with coffee beans differentiation thought high-quality agronomic practices and coffee processing, not necessarily resulting in premium prices for farmers.

Although literature presents a number of positive impacts linked to VSS and certifications in coffee value chains, especially considering coffee farmers "at the bottom of the pyramid" [48], some studies reveal uneven results from region to region. Jena and Grote [47], for instance, observed differences in terms of coffee yield and household income for certified coffee farmers in India, Ethiopia, and Nicaragua, shedding light to role of cooperatives in promoting collective actions and capacity building.

Other issues arise when coffee producers are brought to the center of that discussion: the adoption of certification at the farm level is not always economically viable, once it may bring higher production costs [49]; frequent changes, such as the adoption of new agricultural practices, do not necessarily mean a systemic change toward sustainability [50, 51]; certification is not a synonym of higher prices or better household living and poverty reduction for producers [52, 53].

Despite higher prices at the final market and some changes toward more sustainable production models, certification is not a panacea for sustainability. In that sense, the governance of certification and standards along the value chains plays a central role.

Chain governance agents need to drive more attention to smallholders' inclusion and to support more vulnerable and poorest coffee producers to comply with sustainability standards and develop deep changes toward social and environmental issues [44, 50]. Another governance challenge is related to the producers' awareness of certifications and its meanings, especially for producers in cooperatives or groups [44].

It represents an important alert: collective actions for smallholders' certification may not bring benefits on information sharing, transparency, and administrative competence, compromising its performance in the long run.

The complexity and interactions among the impacts of certifications need to be addressed, to shed light on potential bias or distortions. Impacts on price, for instance, may be related to improvements on quality rather than on the social and environmental aspects of certification itself [44]. Although certification can trigger the development of good agricultural practices and higher levels of assets for producers, the relation may be the other way around: producers already compliant with or close to requirements, or producers already having a minimum level of technical, financial, and structural assets, are generally those who get certified, and not the opposite, which may favor the large-scale producers' adhesion to RA, UTZ, and 4C/GCP certifications [44].

Finally, it is important to consider the interactions and networks for coffee sustainability. Grabs and Carodenutto [54] discuss the role of corporate actors in the governance of sustainable global coffee chain, pointing out benefits but also risks and challenges, such as goal conflict, information asymmetries, and power imbalances. According to Elliott [44], studies also report high levels of dependence on organizations such as NGOs and governmental extension agencies to promote certifications among producers, which sheds light on the need of external assistance and raises questions on the sustainability of certification schemes over time.

The role of the state and public institutions for global value chain upgrading is central. De Marchi and Alford [55] discuss the role of state policies in global value chains, including the coffee one. State regulation is potentially associated with improved social and environmental conditions through the support or requirement of certification schemes. In Brazil, Caldarelli et al. [56] emphasize the importance of public policies to face challenges in Brazilian coffee chain, including efforts to promote quality improvements and social and environmental aspects through voluntary standards and certification schemes.

VSS and sustainable certifications in coffee value chains can emphasize different aspects of sustainability. In general, the adoption of VSS and certifications in coffee value chains brings positive results to the chain and especially to coffee farmers. Promoting product quality, higher revenues, and access to market. Nevertheless, benefits are uneven and not always related to other important indicators, such as household income and coffee yield and producers' empowerment. In that sense, the adoption of VSS and sustainable certifications demands tighter governance. The role of organizations such as cooperatives and governmental agents is crucial to support the adoption of sustainable practices, favor collective actions, and hinder power imbalances between segments, promoting more genuine sustainability in coffee value chains.

5. Coffee production in Brazil

Coffee is produced in more than 60 countries in the coffee belt (between the Tropics of Cancer and Capricorn), but around 70% of the harvest is condensed in four countries: Brazil, Vietnam, Colombia, and Indonesia. Brazil is, by far, the largest producer, with around 33–35% of total production, harvesting both Arabica and Robusta coffee beans [26]. Brazil has been an important (if not the most notable) coffee producer since the eighteenth century, with the commodity having a big role in the history and economy of the country [57].

The first coffee seed came to Brazil in the begin of the eighteenth century in the Northeast, but it was at the end of the century that the plant was introduced in the states of São Paulo, Minas Gerais, Espírito Santo, Paraná, among others, with different types of coffee being planted and modified by genetic engineering [57, 58]. At this time, coffee was planted by the growing high class (centered especially in São Paulo and Rio de Janeiro) in big properties that used slave labor. Later, with the abolition of slavery, labor was due mainly by European immigrants. With the popularization of the brew around the world, coffee had then become the great economic lever for Brazil in the nineteenth century, contributing to the industrialization of the Southeast region. In this century, Brazil was already the largest producer and exporter of the bean [59, 60].

Nowadays, the most prominent regions of production in Brazil are Paraná, São Paulo, Espírito Santo, and Minas Gerais, although, in each one, the coffee had a different role through history, which led to distinct characteristics in production that will be discussed in the following paragraphs [58, 61, 62]. Minas Gerais is the largest producer in Brazil (from 40–50%), with harvest condensed in Sul de Minas, Zona da Mata, and Cerrado Mineiro regions and mainly for Arabica variety. The second largest producer is Espírito Santo (25–30%), harvesting both Arabica and Robusta coffee beans (the state is the biggest producer of Robusta). São Paulo follows, being the third largest producer (Arabica variety) in Brazil (close to 10%). As for Paraná, the state used to be a big and historical producer of Arabica coffee, but, in the last decades, climate adversity has drastically reduced the harvest [63].

The state of São Paulo is one of the oldest producers and the most affected historically and economically by the culture back in the eighteenth and nineteenth centuries. The most important areas of the state are the traditional regions of Alta Paulista and Alta Mogiana, and the relatively younger region of Garça, which begging production after the 1960s decade. The culture in the state applies mostly traditional techniques of cultivation, producing only Arabica beans in small properties [63, 64].

As for Minas Gerais states, the biggest producer, the regions of Zona da Mata and Sul de Minas coffee are also a century and traditional culture characterized by smaller farmers and traditional techniques of cultivation and lower production technology. While in Sul de Minas, only the Arabica variety is harvested, in Zona da Mata, both Arabica and Robusta are planted. As for the region of Cerrado Mineiro, coffee production is relativity new, when producers from São Paulo and Paraná migrated to this region in 1970, due to climate problems in these states and government incentive for a more modern cultivation in Minas Gerais. Due to this, the production of Arabica coffee in Cerrado Mineiro has a higher technological base and is mechanized, a differential from other Brazilian producer regions [57, 61].

In Espírito Santo, the production was initially concentrated in the Arabica variety. In the nineteenth century, the crop had come as a way to occupy the land, organized as big properties focused on the external market. Later, with economical changes, coffee was harvest majorly by small producers, especially in the South of the state, having similar characteristics as Zona da Mata in Minas Gerais [58]. In the North, however, Arabica coffee beans were not adapted for the high temperatures and low altitudes predominant in the area, and with the Programa Federal de Erradicação dos Cafezais (transl. Nation Program of Eradication of Coffee Plantations) in the 1960s decade, most of these crops were annihilated. Producers then started planting Robusta coffee beans, better adapted for the region. The variety had higher productivity and was benefited by the growth of the soluble coffee industry over the years and the expansion of the use of Robusta in blends with Arabica coffee [62].

The difference characteristic among these regions, on the other hand, is of great importance for coffee sustainable production in the context of origin-linked products. The interest in the origin of the coffee seeds began with the second wave of coffee consumption, only gaining more importance in the third wave, with the concept "seed to cup." Coffee producers can then gain competitive advantage and economic benefits by differentiating its products by origin, mainly in the schemes of geographical indications (GIs) [45, 62, 65].

Coffee Geographical Indication Certification was a standard that emerged due to the contributions from representatives of companies, exporters, farmers, and coffee sector stakeholders as a way to increase productivity in farms, to improve market access and the livelihood of coffee famers through sustainable improvement, helping with protection of natural resources and biodiversity [66]. GIs is based on the specific features of products on determined locations, due to a combination of natural resources, traditional local skills and knowledge, and historical and cultural aspects of the origin in question. Producers can then use these different characteristics to add value and promote their products, also protecting the local resources and culture, playing an important role in the sustainable development of local communities [45, 67].

In the economical aspect, GIs have positive impacts by different mechanism, such as providing legal protection for the geographical name of the origin of the product; recognizing the role of primary producers and increase farmer acceptance; boosting competitiveness; positive correlation of GI with intention to pay (premium prices), helping improve farm efficiency and coffee quality; creating new strategies beyond the product (e.g., local ecotourism area [45, 65, 66]. In the coffee scene, the IGs are already commonly used by countries such as Colombia, Indonesia, and Thailand as a way to obtain economic, environmental, and social benefits, such as premium prices, brand value, increase in profit, and decrease in production cost, and improve livelihood of farmers, etc. [66–68].

As for Brazil, the use of GIs in coffee has gained significant importance in the last decades. The protection of GI was determined in Law No. 9.279/1996 in articles 176–182, with the National Institute for Industrial Protection (INPI) responsible for defining procedures for creating GIs and the regulation and control made by the Ministry of Agriculture and Supply (MAPA). According to the law, there are two ways to indicate the geographical region of a product: by Indication of Origin (IP) and Designation of Origin (D.O) [60–62]. The difference between the two of them can be checked in **Table 2**.

In **Table 2**, we can see the difference between the Indication of Origin (IO) and Designation of Origin (D.O), the two kinds of geographic indications find in Brazil. The first one, IO, explicated the name of the origin, functioning based on the notoriety or reputation of the region. As for D.O, works as the very designation of an agricultural or extractive product, whose qualities are intrinsically linked in an exclusive or essential way to the geographical environment.

In 2022, eight IOs and five D.Os for coffee production exist in Brazil [69]. The oldest GI used in the country is the IO for the Cerrado Mineiro region, created in

Indication of origin (IO)	Designation of origin (D.O)
An indication of origin is the geographical name of a country, city, region or locality in its territory, which has become known as a center for the extraction, production or manufacture of a particular product or the provision of a particular service.	Denomination of origin is the geographical name of a country, city, region or locality in its territory, which designates a product or service whose qualities or characteristics are exclusively or essentially due to the geographical environment, including natural and human factors.

Source: Brazil [69], Vieira [60] and Marré & Fonseca [62].

Table 2.
Difference between geographical indications (GIs) in Brazil.

Type of GI	Region (state)	Variety	Creation date	Graphic representation
Designation of origin	Cerrado Mineiro (Minas Gerais)	Arabica	December 2013	
Designation of origin	Mantiqueiras de Minas (Minas Gerais)	Arabica	June 2020	
Designation of origin	Caparaó (Minas Gerais and Espírito Santo)	Arabica	February 2021	
Designation of origin	Montanhas do Espírito Santo (Espírito Santo)	Arabica	May 2021	
Designation of Origin	Matas de Rondônia (Rondônia)	Robusta	June 2021	
Indication of origin	Cerrado Mineiro (Minas Gerais)	Arabica	April 2005	No representation
Indication of origin	Norte Pioneiro do Paraná (Paraná)	Arabica	September 2012	
Indication of origin	Alta Mogina (São Paulo)	Arabica	September 2013	
Indication of origin	Região de Pinhal (São Paulo)	Arabica	July 2016	
Indication of origin	Oeste do Paraná (Paraná)	Arabica	July 2017	
Indication of origin	Oeste da Bahia (Bahia)	Arabica	May 2019	

Type of GI	Region (state)	Variety	Creation date	Graphic representation
Indication of origin	Campo das Vertentes (Minas Gerais)	Arabica	November 2020	CAMPO DAS VERTENTES
Indication of origin	Matas de Minas (Minas Gerais)	Arabica	December 2020	No representation
Indication of origin	Espírito Santo (Espírito Santo)	Robusta	May 2021	Café Conilon Espírito Santo Indicação de Procedência

Source: Brazil [69].

Table 3.
Types of geographical indications (GIs) for coffee in Brazil.

2005, and according to Almeida and Tabaral [61], it's the first region in the world to issue a D.O. seal for green coffee as well for roasting coffee in 2013. Other GIs in Brazil are issued in the main producing regions of coffee of Minas Gerais, São Paulo, Paraná, Espírito Santo, Rondônia, and Bahia states, accounting for more than 400 cities around the country [69].

A crescent investment for producers in the last years is regarding the Robusta Beans that are achieving recognition in the global market. Thus, the last GI appointed by the government is the IOs for Espírito Santo and Matas de Rondônia.

Figure 1.
GI of coffee in Brazil. Source: Adapted from Brazil [69].

For this last IO, there is a crest adoption of the denomination "Amazonian Robustas," which also reflects the concerns of coffee producers with sustainability [69]. All of the IOs and D.Os can be observed in **Table 3** and **Figure 1**.

In **Table 3** and **Figure 1**, the multiple IOs and D.Os of coffee in Brazil are shown. **Table 3** presents the variety of coffee in question, as well as the state, creation date, and the logo of each geographic indication. In **Figure 1**, the same IO and D.O are shown in the map of Brazil. As we can see, these geographical indications are present in six states, but are concentrated in Minas Gerais state, the largest producer in Brazil.

6. Final remarks

The differentiation strategy aiming value creation for coffee farmers in Brazil has been in place since the deregulation of coffee market in mid-1990s [70]. Industry played an important role to define quality standards through the Brazilian Coffee Quality Program since 1989 with continual enrichment aiming to match the growing interest of the consumers toward the coffee origin and quality. The private and public prizes rewarding famers and roasters for good practices have been another salient institutional tool to achieve and enhance quality and sustainable practices along the coffee chain.

Nevertheless, the VSS adoption and GI's creation have been modern strategies for quality and suitability achievement as demonstrated in this chapter. Brazil, as the largest global coffee producer, has also a great potential regarding GI strategies, which can lead to differentiation strategies and economic benefits for small farmers, also contributing to sustainable production and valorization of the cultural environmental of these regions. However, public and private action should consider economic and social barriers to achieve the VSS and VI's protocols, developing means to foster, maintain, and enhance a quality and sustainability mind set along the coffee chain [60, 65, 71].

The coffee value chain has great importance in the agribusiness, involving a huge number of actors from its production to its consumption, and Brazil has a huge part in this as the largest producer and second largest consumer. Thus, in the context of sustainability in the GVC, it's important to look more thoroughly in the aspects of the Brazilian coffee scenario. Around the world, the sector is already considered a pioneer in the adoption of VSSs, in particular, "private" and multistakeholder initiatives, such as Fair Trade and Organic certifications, which are also applied in Brazil.

Yet, these VSSs are mainly driven by the consumer ends, and there has been contrasting evidence of the real effectiveness of these standards in the incomes and livelihood of producers, thus presenting a possible challenge in the sustainability of the chain. What is known is that these standards may have different effects depending on the country studied.

Author details

Laleska Rossi Moda[1], Eduardo Eugênio Spers[1*], Luciana Florêncio de Almeida[2] and Sandra Mara de Alencar Schiavi[3]

1 Departament of Economics, Administration and Sociology (LES), Luiz de Queiroz College of Agriculture (ESALQ), University of São Paulo (USP), Piracicaba, SP, Brazil

2 Professional Master's in Consumer Behavior (MPCC) – ESPM, Brazil

3 State University of Maringá (UEM), Maringá, PR, Brazil

*Address all correspondence to: edespers@usp.br

IntechOpen

References

[1] Baumgärtner S, Quaas M. What is sustainability economics? Ecological Economics. 2010;**69**:445-450. DOI: 10.1016/j.ecolecon.2009.11.019

[2] Lee MT, Raschke RL, Krishen AS. Signaling green! Firm ESG signals in an interconnected environment that promote brand valuation. Journal of Business Research. 2022;**138**:1-11. DOI: 10.1016/j.jbusres.2021.08.061

[3] Bahadur Kc K, Dias GM, Veeramani A, Swanton CJ, Fraser D, Steinke D, et al. When too much isn't enough: Does current food production meet global nutritional needs? PLoS One. 2018;**13**:e0205683. DOI: 10.1371/journal. pone.0205683

[4] Hannah L, Roehrdanz PR, Krishna Bahadur KC, Fraser EDG, Donatti CI, Saenz L, et al. The environmental consequences of climate-driven agricultural frontiers. PLoS One. 2020;**15**:e0228305. DOI: 10.1371/journal. pone.0228305

[5] Fraser E, Legwegoh AKCK, CoDyre M, Dias G, Hazen S, Johnson R, et al. Biotechnology or organic? Extensive or intensive? Global or local? A critical review of potential pathways to resolve the global food crisis. Trends in Food Science and Technology. 2016;**48**:78-87. DOI: 10.1016/j. tifs.2015.11.006

[6] Bager SL, Lambin EF. Sustainability strategies by companies in the global coffee sector. Business Strategy and the Environment. 2020;**29**:3555-3570. DOI: 10.1002/bse.2596

[7] Panhuysen S, Pierrot J. Coffee Barometer: 2014. Zeitschrift für Analytische Chemie. 2014;**28**:1-15

[8] Kuhlman T, Farrington J. What is sustainability? Sustainability. 2010;**2**:3436-3448. DOI: 10.3390/su2113436

[9] Rosen MA, Dincer I, Hacatoglu K. Sustainability. In: In: Wexler P., editor. Encyclopedia of Toxicology. 3rd ed. Oshawa: Academic Press; 2014. pp. 442-447. DOI: 10.1016/ B978-0-12-386454-3.01046-0

[10] Visser W. Our Common Future ('The Brundtland Report') World Commission on Environment and Development (1987). In: The Top 50 Sustainability Books. London: Routledge. 2017; p. 52-55

[11] Kudo S, Mino T. Framing in sustainability science. In: Mino T, Kudo S, editors. Framing in Sustainability Science. Science for Sustainable Societies. Singapore: Springer; 2020. DOI: 10.1007/978-981-13-9061-6_1

[12] Goodland R. The concept of environmental sustainability. Annual Review of Ecology and Systematics.1995;**26**:1-24. DOI: 10.1146/annurev.es.26.110195.000245

[13] Hajian M, Jangchi KS. Evolution of the concept of sustainability. From Brundtland report to sustainable development goals. In: Hussain C M, Velasco-Munoz JF, editors. Sustainable Resource Management: Modern Approaches and Contexts. Elsevier. Sustainable Resource Management: Modern Approaches and Contexts. Cambridge: Elsevier; 2021. p. 1-24. DOI: 10.1016/B978-0-12-824342-8.00018-3

[14] Kori E, Gondo T. Environmental sustainability: Reality, fantasy or fallacy. In: Proceedings of 2nd International Conference on Environment and BioScience. 28-29 September 2012; Phnom Penh - Cambodia. Singapore: IACSIT Press; 2012. p. 44

[15] Murali Krishna IV, Manickam V. Sustainable Development. In: Murali Krishna IV, Manickam V. Environmental

Management: Science and Engineering for Industry. Oxford: Butterworth-Heinemann; 2017. p. 5-21. DOI: 10.1016/B978-0-12-811989-1.00002-6

[16] United Nations. The 17 Sustainble development goals. In: Department of Economic and Social Affairs. 2020

[17] United Nations. The sustainable development goals report 2020. 2020. Available from: https://unstats.un.org/sdgs/report/2020/The-Sustainable-Development-Goals-Report-2020.pdf [Accessed: 2022-05-23]

[18] Anshari M, Almunawar MN, Masri M, Hamdan M. Digital marketplace and FinTech to support agriculture sustainability. Energy Procedia. 2019;**156**:234-238. DOI: 10.1016/j.egypro.2018.11.134

[19] Stagnari F, Maggio A, Galieni A, Pisante M. Multiple benefits of legumes for agriculture sustainability: An overview. In chemical and biological technologies. Agriculture. 2017;**4**:1-13. DOI: 10.1186/s40538-016-0085-1

[20] Giovannucci D, Ponte S. Standards as a new form of social contract? Sustainability initiatives in the coffee industry. Food Policy. 2005;**30**:284-301. DOI: 10.1016/j.foodpol.2005.05.007

[21] Grabs J, Ponte S. The evolution of power in the global coffee value chain and production network. Journal of Economic Geography. 2019;**19**:803-828. DOI: 10.1093/jeg/lbz008

[22] Takahashi R. How to stimulate environmentally friendly consumption: Evidence from a nationwide social experiment in Japan to promote eco-friendly coffee. Ecological Economics. 2021;**186**:107082. DOI: 10.1016/j.ecolecon.2021.107082

[23] USDA. United States Department of Agriculture. Market and Trade Data. Foreign Agricultural Service. Custom Query. 2022. Available from: https://apps.fas.usda.gov/psdonline/app/index.html#/app/advQuery [Accessed 2022-01-30]

[24] CECAFE, Conselho dos Exportadores de Café do Brasil. Relatório mensal dezembro de 2021. 2022. Available from: www.cecafe.com.br [Acessed 2022-02-20]

[25] Fischer EF. Quality and inequality: Taste, value, and power in the third wave coffee market. In: MPIfG Discussion Paper. 2017. DOI: http://hdl.handle.net/10419/156227%0A

[26] ICO. The value of Coffee: Sustainability, Inclusiveness, and Resilience of the Coffee Global Value Chain. 2020b. Available from: https://www.ico.org/documents/cy2020-21/ed-2357e-cdr-2020.pdf [Accessed 2021-10-03]

[27] ICO - International Coffee Organization. World Coffee Consumption. 2022. Available from: https://www.ico.org/ [Accessed: 2022-05-23]

[28] Grabs J. Assessing the institutionalization of private sustainability governance in a changing coffee sector. Regulation and Governance. 2020;**14**:362-387. DOI: 10.1111/rego.12212

[29] Manning S, Boons F, von Hagen O, Reinecke J. National contexts matter: The co-evolution of sustainability standards in global value chains. Ecological Economics. 2012;**83**:197-209. DOI: 10.1016/J.ECOLECON.2011.08.029

[30] World Bank. World Development Report 2020 - Global Value Chains: Trading for Development. 2022. Available from: https://www.worldbank.org/en/publication/wdr2020 [Accessed: 2022-05-23]

[31] Economist Intelligence Unit. An Eco-wakening - Measuring global

awareness, engagement and action for nature. 2021. p.1-50. Available from: https://files.worldwildlife.org/wwfcmsprod/files/Publication/file/93ts5bhvyq_An_EcoWakening_Measuring_awareness__engagement_and_action_for_nature_FINAL_MAY_2021.pdf [Accessed: 2022-05-23]

[32] Abdu N, Mutuku J. Willingness to pay for socially responsible products: A meta–analysis of coffee ecolabelling. Heliyon. 2021;**7**(6):e07043. DOI: 10.1016/j.heliyon.2021.e07043

[33] Akoyi KT, Maertens M. Walk the talk: Private sustainability standards in the Ugandan coffee sector. Journal of Development Studies. 2018;**54**:1792-1818. DOI: 10.1080/00220388. 2017.1327663

[34] Bacon CM, Méndez VE, Gómez MEF, Stuart D, Flores SRD. Are sustainable coffee certifications enough to secure farmer livelihoods? The millenium development goals and Nicaragua's fair trade cooperatives. Globalizations. 2008;**5**:259-274. DOI: 10.1080/14747730802057688

[35] Reinecke J, Manning S, von Hagen O. The emergence of a standards market: Multiplicity of sustainability standards in the global coffee industry. Organization Studies. 2012;**33**:791-814. DOI: 10.1177/0170840612443629

[36] Dietz T, Auffenberg J, Estrella Chong A, Grabs J, Kilian B. The voluntary coffee standard index (VOCSI). Developing a composite index to assess and compare the strength of mainstream voluntary sustainability standards in the global coffee industry. Ecological Economics. 2018;**150**:72-87. DOI: 10.1016/j.ecolecon.2018.03.026

[37] Panhuysen S, Pierrot J. Coffee Barometer: 2020. Coffee Collective 2020. Available from: https://coffeebarometer.org/ [Accessed: 2022-05-23]

[38] Global Coffee Platform. Sustainable Coffee Purchases Snapshot 2019 & 2020. 2021

[39] Henson S, Humphrey J. The impacts of private food safety standards on the food chain and on public standard-setting processes. Joint FAO/WHO Food Standards. 2009. Available from: https://www.fao.org/3/i1132e/i1132e.pdf [Accessed 2018-05-04]

[40] Giovannucci D, Ponte S. Standards as a new form of social contract? Sustainability initiatives in the coffee industry. Food Policy. 2005;**30**:284-301. DOI: 10.1016/j.foodpol.2005.05.007 Accessed: 2022-01-29

[41] Vanderhaegen K, Akoyi KT, Dekoninck W, Jocqué R, Muys B, Verbist B, et al. Do private coffee standards 'walk the talk' in improving socio-economic and environmental sustainability? Global Environmental Change. 2018;**51**:1-9. DOI: 10.1016/j.gloenvcha.2018.04.014

[42] Piao RS, Fonseca L, Januário EC, Saes MSM, Almeida LF. The adoption of voluntary sustainability standards (VSS) and value chain upgrading in the Brazilian coffee production context. Journal of Rural Studies. 2019;**71**:13-22. DOI: 10.1016/j.jrurstud.2019.09.007

[43] Grabs J. Assessing the institutionalization of private sustainability governance ina changing coffee sector. Regulation & Governance. 2018;**14**:362-387. DOI: 10.1111/rego.12212

[44] Elliott KA. What Are we Getting from Voluntary Sustainability Standards for Coffee? CDG Policy Paper 129. Washington, DC: Center for Global Development; 2018. Available from: https://www.cgdev.org/publication/what-are-we-getting-voluntary-sustainability-standards-coffee Accessed 2020-09-09

[45] Vandecandelaere E, Teyssier C, Barjolle D, Fournier S, Beucherie O, Jeanneaux P. Strengthening sustainable food systems through geographical indications: Evidence from 9 worldwide case studies. Journal of Sustainability Research. 2020;**2**:200031. DOI: 10.20900/jsr20200031

[46] Cabrera LC, Caldarelli CE, Camara MRG. Mapping collaboration in international coffee certification research. Scientometrics. 2020;**124**:2597-2618. DOI: 10.1007/s11192-020-03549-8

[47] Jena PR, Grote U. Do certification schemes enhance coffee yields and household income? Lessons learned across continents. Front. Sustain. Food Syst. 2022;**5**:716904. DOI: 10.3389/fsufs.2021.716904

[48] Karki SK, Jena PR, Grote U. Fair trade certification and livelihoods: A panel data analysis of coffee-growing households in India. Agricultural Economics. 2016;**45**:436-458. DOI: 10.1017/age.2016.3

[49] Cabrera, LC. Caldarelli, CE. Viabilidade econômica de certificações de café para produtores brasileiros. Revista de Política Agrícola 2021; 30: 64-76. Available from: https://seer.sede.embrapa.br/index.php/RPA/article/view/1651/pdf [Accessed 2022-01-08]

[50] Glasbergen P. Smallholders do not eat certificates. Ecological Economics. 2018;**147**:243-252. DOI: 10.1016/j.ecolecon.2018.01.023

[51] Jena PR, Grote U. Fairtrade certification and livelihood impacts on small-scale coffee producers in a tribal community of India. Applied Economic Perspectives and Policy. 2017;**39**:87-110. DOI: 10.1093/aepp/ppw006

[52] Jena PR, Chichaibelu BB, Stellmacher T, Grote U. The impact of coffee certification on small-scale producers' livelihoods: A case study from the Jimma zone. Ethiopia.

Agricultural Economics. 2012;**43**(4): 429-440. DOI: 10.1111/j.1574-0862.2012.00594.x

[53] Chiputwa B, Spielman DJ, Qaim M. Food standards, certification, and poverty among coffee farmers in Uganda. World Development. 2015;**66**:400-412. DOI: 10.1016/j.worlddev.2014.09.006

[54] Grabs J, Carodenuto SL. Traders as sustainability governance actors in global food supply chains: A research agenda. Business Strategy and the Environment. 2021;**30**:1314-1332. DOI: 10.1002/bse.2686

[55] De Marchi V, Alford M. State policies and upgrading in global value chains: A systematic literature review. Journal of International Business Policy. 2022;**5**:88-111. DOI: /10.1057/s42214-021-00107-8

[56] Caldarelli CE, Gilio L, Zilberman D. The coffee market in Brazil: Challenges and policy guidelines. Revista de Economia. 2018;**39**:1-21. DOI: 10.5380/re.v39i69.67891

[57] Santana GHDS, Geográfica I. do café da região do cerrado mineiro: formação, consolidação e perspectivas. In: Trabalho de Conclusão de Curso (Graduação em Geografia). Uberlândia: Universidade Federal de Uberlândia; 2020

[58] Melo JR, Silva NFM, Siqueira Nunes NM. Café: origem e contribuição para a economia do Brasil. Revista Científica Interdisciplinar. 2018;**3**:15-24

[59] Simão J, Souza MA, Silva MV, Zanúncio Junior JS. Denominação de origem Caparaó para Café Arábica. Incaper em Revista. 2021;**11**-12:49-60. DOI: 10.54682/ier.v11e12-p49-60

[60] Vieira ACP, Pellin V, Bruch KL, Locatelli L. (2019). Desenvolvimento regional e indicações geográficas de café no Brasil: perspectivas pós-registro.

In: Vieira ACP, Lourenzani AEBS, Bruch Kelly L, Locatelli L, Gaspar LCM (org.). Indicações geográficas, signos coletivos e o desenvolvimento local/regional. Erechim: Deviant, 2019. p. 171-198

[61] Almeida LF, Tarabal J. Cerrado Mineiro region designation of origin: Internationalization strategy. In: Almeida LFD, Spers EE. Coffee Consumption and Industry Strategies in Brazil. Cambridge: Elsevier; 2020. p. 189-202. DOI: 10.1016/B978-0-12-814721-4.00008-1

[62] Marré WB, Fonseca AFA. Indicação de Procedência (IP) Espírito Santo para o Café Conilon (Coffea canephora). Incaper em Revista. 2021;**11-12**:99-107. DOI: 10.54682/ier.v11e12-p99-107

[63] Conab. Companhia Nacional de Abastecimento. In: Acompanhamento da safra brasileira de café - Safra 2022 – Primeiro levantamento. Brasília: CONAB; 2022. pp. 1-60

[64] São Paulo. Government of São Paulo. Agricultura de SP apoia trabalhos de toda a cadeia de produção de café. 2020. Retrieved from: https://www.saopaulo.sp.gov.br/ultimas-noticias/agricultura-de-sp-apoia-trabalhos-de-toda-a-cadeia-de-producao-de-cafe/#:~:text=A%20cafeicultura%20%C3%A9%20um%20tema,a%20import%C3%A2ncia%20dessa%20atividade%20agr%C3%ADcola

[65] Lourenzani AEBS, Watanabe K, Pigatto GAS, Pereira MEG. What fills your cup of coffee? The potential of geographical indication for family farmers' market access. In: Almeida LFD, Spers EE. Coffee Consumption and Industry Strategies in Brazil. Cambridge: Elsevier; 2020. p. 149-165. DOI: 10.1016/B978-0-12-814721-4.00014-7

[66] Pratiwi LPK, Budiasa M, Yudiarini N. The role of the geographic indication certification of arabic coffee as an effort of local farmers. International Journal of Research-GRANTHAALAYAH. 2021;**9**:330-338. DOI: 10.29121/granthaalayah.v9.i2.2021.3102

[67] Nizam D, Tatari MF. Rural revitalization through territorial distinctiveness: The use of geographical indications in Turkey. Journal of Rural Studies. 2020; in press: 1-11. DOI: 10.1016/j.jrurstud.2020.07.002

[68] Lilavanichakul A. The economic impact of Arabica coffee farmers' participation in geographical indication in northern Highland of Thailand. Journal of Rural Problems. 2020;**56**: 124-131. DOI: 10.7310/arfe.56.124

[69] Brazil. Ministry of Agriculture and Supply.Pela primeira vez, seminário reúne representantes de todas as Indicações Geográficas de Café registradas no Brasil. 2022. Access feb, 2022. Retrieved from: https://www.gov.br/agricultura/pt-br/assuntos/noticias/1o-seminario-de-indicacoes-geograficas-de-cafe-reune-todas-as-igs-registradas-no-brasil

[70] Saes MSM, Escudeiro FH, da Silva CL. Differentiation strategy in the Brazilian coffee market. Review of Business Management. 2006;**8**(21):24-32

[71] Ruiz XFQ, Nigmann T, Schreiber C, Neilson J. Collective action milieus and governance structures of protected geographical indications for coffee in Colombia, Thailand and Indonesia. International Journal of the Commons. 2020;**14**:329-343. DOI: 10.5334/ijc.1007